# Lecture Notes in Physics

Edited by H. Araki, Kyoto, J. Ehlers, München, K. Hepp, Zürich
R. Kippenhahn, München, D. Ruelle, Bures-sur-Yvette
H. A. Weidenmüller, Heidelberg, J. Wess, Karlsruhe and J. Zittartz, Köln
Managing Editor: W. Beiglböck

## 338

V. Privman
N.M. Švrakić

W0192929

# Directed Models of Polymers, Interfaces, and Clusters: Scaling and Finite-Size Properties

Springer-Verlag Berlin Heidelberg GmbH

**Authors**

V. Privman
N. M. Švrakić*
Department of Physics, Clarkson University
Potsdam, NY 13676, USA

* on leave of absence from
Institute of Physics
P.O. Box 57, 11001 Belgrade, Yugoslavia

ISBN 978-3-662-13717-8       ISBN 978-3-540-48120-1 (eBook)
DOI 10.1007/978-3-540-48120-1

Originally published by Springer-Verlag Berlin Heidelberg New York in 1989
Softcover reprint of the hardcover 1st edition 1989

# PREFACE

These *Lecture Notes* provide a detailed introduction to, as well as an exposition of, research results for various, mostly two-dimensional, models of directed walks, interfaces, wetting, surface adsorption (of polymers), stacks, compact clusters (lattice animals), etc. The unifying feature of these models is that in most cases they can be solved analytically. The methods used include transfer matrices, generating functions, recurrence relations, and difference equations, and in some cases involve utilization of less familiar mathematical techniques such as continued fractions and $q$-series.

We emphasize an overall view of what can be learned generally of the statistical mechanics of anisotropic systems, including phenomena near surfaces, by studying the solvable models. Thus, the concept of scaling and, where known, finite-size scaling properties are elucidated. Scaling and statistical mechanics of anisotropic systems in general are active research topics. We hope that our monograph will provide a comprehensive survey of exact model results in this field.

While we do not give an exhaustive list of references, we provide selected literature lists at the end of each chapter, including suggested general review articles and books. The specific selection of topics and details of presentation have been influenced in many cases by our own research interests and experience. We wish to thank our colleagues, D.B. Abraham, M. Barma, G. Bilalbegović, M.C.T.P. Carvalho Bartelt, F. Family, M.E. Fisher, G. Forgacs, H.L. Frisch, M.L. Glasser, S. Redner, J. Rudnick, L.S. Schulman, A.M. Szpilka, and R.K.P. Zia for rewarding interactions and collaboration, which contributed to our understanding of the physics of anisotropic systems. Most of our research results reported in this monograph have come from work supported in part by the United States National Science Foundation (under grant DMR-86-01208 to VP), and by the Donors of the Petroleum Research Fund, administered by the American Chemical Society (under grant ACS-PRF-18175-G6 to VP). This financial assistance is gratefully acknowledged.

Potsdam, New York, March 1989            Vladimir Privman

Nenad M. Švrakić

# Table of Contents

# I. INTRODUCTION

During the past several decades, geometrical lattice models such as random walks, strings, clusters, etc., have played an important role in Statistical Mechanics and in particular, in the theory of critical phenomena. Lattice models, in spite of their apparent simplicity, can be successfully used in the studies of real physical systems, e.g., polymers and interfaces, to describe conformational and growth features, adsorption, wetting, etc. In this monograph we consider several solvable anisotropic lattice models and analyze their scaling behavior and in some cases, finite-size properties. We focus on simple mostly two-dimensional lattice models and emphasize generic features characterizing their scaling properties.

In order to make our models exactly solvable, we impose "microscopic" restrictions such as partial or full *directedness* by, e.g., disallowing certain steps in a random walk, or *compactness* for cluster models (lattice animals). The resulting anisotropic models are in many instances exactly solvable by methods including generating functions, transfer matrices, continued fractions, etc. Some of the less familiar mathematical techniques are briefly reviewed in appropriate sections.

The format of this monograph is as follows. In Chapter II, we study directed random walk models of polymer chain conformations. This is probably the simplest model that is solvable in full detail on a $d$-dimensional lattice, and its scaling properties can be derived exactly. By assigning statistical weight to each turn in a walk, we can also describe the rod-to-coil transition of linear polymers and analyze the appropriate scaling properties. Finally, we consider directed walks on finite-size lattices and derive some finite-size scaling results.

Chapter III is devoted to the solid-on-solid models (SOS) of line interfaces. Geometrically, such interfaces are identical to directed random walks. However, different interface configurations are assigned different energy. SOS models have been extensively used to describe depinning of interfaces in the presence of various substrate potentials. The most frequently used method of solution is the continuous limit (differential equation) approximation, and we briefly review the results of such studies. However, in the presence of long range substrate potentials it is important to retain lattice (discrete) description, since the continuous limit may not always be justified. We briefly survey mathematical methods of solution of the appropriate difference equations and describe exact results for wetting transitions in systems with short and long range substrate potentials. Finally, we analyze finite-size properties of fluctuating SOS interfaces, both unbound and in the regime near the wetting transition.

Directed walk models of polymers can be used to study the behavior of a single polymer chain near an attractive surface. We describe such a study in Chapter IV. By assigning proper statistical weights to bulk and surface steps one can solve the model exactly (by the transfer matrix method) and analyze adsorption-desorption transition at impenetrable and penetrable surfaces in two and three dimensional systems.

Chapter V is devoted to the two dimensional lattice models of compact clusters (lattice animals), which in many cases can be solved exactly for the generating functions of the cluster numbers. The mathematical methods for solving the appropriate difference equations are sometimes rather sophisticated, including, e.g., continued fractions and $q$-series. A brief review of these techniques is given. Finally, a solvable model of finite-size scaling properties of the partially directed compact lattice animals is presented. In Chapter VI, we give a brief summary of the results and mention some open problems.

## II. DIRECTED WALK MODELS OF POLYMER CONFORMATIONS

In this chapter we consider directed self-avoiding walk (DSAW) models and their relevance for description of conformational properties of polymers. First, in Section A, several random walk models are defined and appropriate physical quantities introduced. Section B presents the generating function formalism for solution of DSAW problems. In Section C we analyze in detail the rod-to-coil transition of linear polymers and discuss the exact solution of this problem. Section D is devoted to scaling properties of polymers in confined geometries. Selected literature is listed in Section E.

### A. Definition of the model

Linear polymers are probably the simplest physical systems that can be studied in the framework of random walk models. They are long, chain-like molecules formed by repetitions of a basic unit or segment. These monomer units are typically connected by carbon-carbon covalent bonds, and a single polymer chain can consist of any number of basic units. More importantly, a long polymer chain is *flexible*, i.e., it can assume different geometric configurations. Experimental evidence shows that physical properties of polymers depend strongly on the statistics of their conformations.

In order to study the average geometrical features of polymers (their sizes, shapes, etc.), a chain of $N$ monomers is represented by a broken line consisting of $N$ segments. For mathematical convenience, the configurations of such a line or *walk* are considered on a regular $d$-dimensional lattice. Thus, in its simplest form, this approach models polymer configurations by ordinary *random walks*, i.e., a succession of $N$ steps, starting from some origin and reaching an arbitrary end-point. The polymer of $N$ monomer segments is represented by an $N$-step walk. In Figs. 1 (a)-(d), several types of walks on a $d = 2$ square lattice, with successive steps labeled by numbers, are shown for illustration. Depending on the physical situation, one may impose additional restrictions such as self-avoidance, directedness, etc., and we will return to these later. Presently, let us introduce some notation commonly used for all types of walks.

One of the quantities of interest in characterizing the random walks is the number, $c_N$, of different walks of length $N$. Specifically, for large values of $N$, the total number of walks is believed to grow according to

$$c_N = C \mu^N N^{\gamma-1}(1 + ...), \tag{2.1}$$

where the exponent $\gamma$ is *universal*, i.e., independent of the microscopic details such as the type of the underlying lattice (square, triangular, etc.). However, $\gamma$ may depend on the dimensionality, $d$, of the lattice and other global features like self-avoidance, directedness, etc. The growth parameter $\mu$ in (2.1) is the connective constant, or cardinality, and, as we will see, plays the role analogous to that of the inverse critical temperature in phase transition models. Finally, the constant $C$ in (2.1) is a nonuniversal coefficient, while the dots indicate higher-order terms which vanish as $N \to \infty$.

Another characteristic quantity is the mean square end-to-end distance, $\langle R_N^2 \rangle$, which has the limiting behavior

$$\langle R_N^2 \rangle = A N^{2\nu}(1 + ...), \tag{2.2}$$

where the exponent $\nu$ is also universal and characterizes the leading growth rate of the polymer size as $N \to \infty$. We emphasize that the scaling relations (2.1) and (2.2) are quite general.

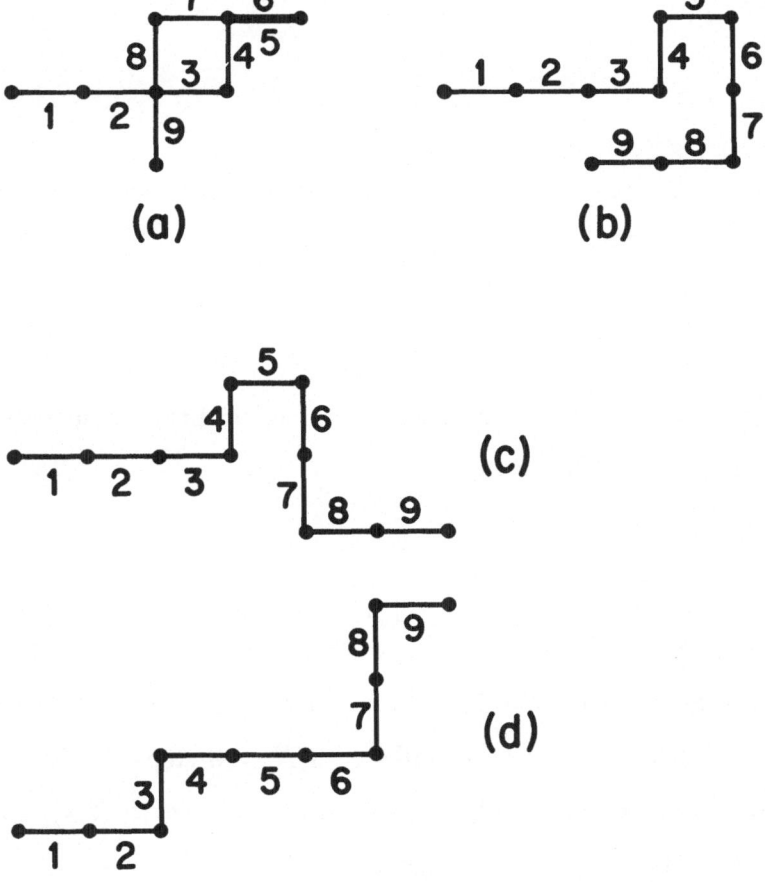

**Figure 1**

Nine-step walks on the square lattice: (a) Gaussian (non self-avoiding); (b) self-avoiding; (c) partially directed SAW ($+X$ and $\pm Y$ steps); (d) fully directed SAW ($+X$ and $+Y$ steps only).

The knowledge of quantities like $c_N$ and $\langle R_N^2 \rangle$ enables one to study geometrical and statistical properties of polymer chains. For example, the number $c_N$ yields information on the conformational entropy of the chain. In addition, one can establish the analogy between $c_N$, $\langle R_N^2 \rangle$, etc., and the well-known quantities in the critical phenomena theory. Specifically, the generating function for the numbers $c_N$,

$$\chi(z) \equiv \sum_{N=0}^{\infty} c_N z^N, \qquad (2.3)$$

where $z$ is the "fugacity", is analogous to the susceptibility. (We take $c_0 \equiv 1$.) Similarly, the quantity

$$\xi^2(z) \equiv \frac{\sum \langle R_N^2 \rangle c_N z^N}{\sum c_N z^N}, \qquad (2.4)$$

plays the role of the correlation length. Near criticality, the susceptibility and the correlation length behave as

$$\chi \sim (z_c - z)^{-\gamma}, \quad \text{and} \quad \xi \sim (z_c - z)^{-\nu}. \qquad (2.5)$$

These relations follow from (2.1)-(2.4), with the critical fugacity value $z_c \equiv \mu^{-1}$. For technical reasons it is sometimes more convenient to put $c_0 \equiv 0$ in definitions like (2.3), i.e., not to include the walks with $N = 0$ steps. This has an effect of modifying the precise form of, e.g., the correlation length (2.4); however, the critical behavior near $z_c$ remains unaffected. We use this convention in Section B, for instance.

In the case of lattice walks with no restrictions (Gaussian random walks), relations (2.1)-(2.2) take simple forms

$$c_N \equiv \mu_0^N, \quad \text{and} \quad \langle R_N^2 \rangle \equiv a^2 N. \qquad (2.6)$$

In these expressions, $\mu_0$ is the coordination number of the lattice while $a$ is the lattice spacing (step length). In the general case, $\mu$ is sometimes also termed the *effective* coordination number. Note that (2.6) implies $\gamma = 1$ and $\nu = 1/2$, in the case of Gaussian walks, independent of the dimensionality of space, which is a typical mean-field property. The relations (2.6) can be derived easily if we recall that the probability for the walk to proceed from some point on the lattice in any direction is equal to the inverse of the coordination number, $1/\mu_0$, and that the steps are statistically independent. Using definitions (2.3) and (2.4), with (2.6), we obtain

$$\chi(z) = \frac{1}{1 - z\mu_0}, \quad \text{and} \quad \xi^2(z) = \frac{a^2 z \mu_0}{1 - z\mu_0}, \tag{2.7}$$

which is of the form (2.5) with $z_c = 1/\mu_0$.

The Gaussian walk has simple properties because of its Markovian character, i.e., the successive steps are completely independent (uncorrelated). Such walk is illustrated in Fig. 1 (a). A more realistic model for conformational properties of polymers is obtained if the walk is subject to the restriction that it may pass only once through any lattice point. This restriction is appropriate for polymers in solutions, for instance. Such walk is termed non-self-intersecting or *self-avoiding* (SAW). Examples of SAWs are shown in Fig. 1 (b)-(d). Enumeration of SAW on various lattices is a much studied problem (in the literature on the subject this problem is also termed the excluded volume problem), but so far only few exact results have been conjectured, mostly in $2d$. However, there is a large amount of numerical data obtained by exact counting of walks for finite $N$ (usually up to $N \sim 20$), or by Monte Carlo methods where walks of up to several hundred steps have been generated. These numerical studies show that $c_N$ and $\langle R_N^2 \rangle$ for SAW behave according to (2.1) and (2.2), but that the exponents $\gamma$ and $\nu$ depend on the dimensionality, $d$, of the lattice, and that $\mu_c$ is somewhat less than the lattice coordination number (for $d > 1$). Specifically, for $d = 1$ the exponents can be calculated exactly, giving $\gamma_1 = 1$ and $\nu_1 = 1$. For $d = 2$, the exact values have been conjectured, $\gamma_2 = 43/32$ and $\nu_2 = 3/4$, while for $d = 3$, numerical studies suggest $\gamma_3 \sim 1.16$ and $\nu_3 \sim 0.59$. Note that with increasing dimensionality of the lattice the exponents approach their "ideal" (Gaussian) values $\gamma = 1$ and $\nu = 1/2$. Indeed, it has been shown that for $d > 4$, and up to logarithmic correction factors in $d = 4$, the SAW is characterized by the same exponents as the Gaussian walk.

In addition to self-avoidance, one can introduce further constraints regarding the direction of the walk by, e.g., preventing certain types of steps. With this restriction, a *directed* SAW is obtained. Figs. 1 (c)-(d) show two DSAWs on the square lattice: Fig. 1 (c) presents a DSAW in which negative-$X$ steps are not allowed, while in Fig. 1 (d) both negative-$X$ and negative-$Y$ steps are excluded. Such walks are of interest because they can be *solved exactly* in many instances; they also illustrate how a "microscopic" bias affects the exponents entering the scaling forms (2.1) and (2.2). In order to characterize DSAW, the preferred "directed" space axis must be identified. Because of the preferred direction of the walk, the mean square end-to-end distances $\langle R^2_{N\parallel} \rangle$ and $\langle R^2_{N\perp} \rangle$, corresponding, respectively, to displacements parallel and orthogonal to the directed axis, are expected to scale with different exponents,

$$\langle R^2_{N\parallel} \rangle = A_{\parallel} N^{2\nu_{\parallel}}(1 + ...), \qquad (2.8)$$

and

$$\langle R^2_{N\perp} \rangle = A_{\perp} N^{2\nu_{\perp}}(1 + ...). \qquad (2.9)$$

Relation (2.1), however, remains unchanged. The exponents $\nu_{\parallel}$ and $\nu_{\perp}$ characterize the correlations parallel and transverse to the directed axis. Note that the axis of directedness of the walk does not have to coincide with the principal lattice axis. It turns out that, typically, the correlations parallel to the directed axis are effectively one-dimensional, and $\nu_{\parallel} = 1$; the orthogonal steps are effectively uncorrelated, and one finds the Gaussian exponent $\nu_{\perp} = 1/2$. These conclusions, which also hold for higher dimensional lattices, are confirmed by exact calculations described in the next section.

## B. Generating function formalism and exact results

During the past several decades a number of methods, both numerical and analytical, have been developed for the solution of the random walk problems. Among the numerical methods, we already mentioned the direct counting and Monte Carlo simulations. The analytical techniques include field theoretical methods, the transfer matrix approach, and various versions of the generating function method. The transfer matrix method is useful in the studies of interfaces, and will be described in Chapter III. In this chapter we focus on the generating function approach, which is one of the simplest techniques, and is also applicable to other problems that will be considered later.

The generating function technique is frequently used for the solution of random walk problems and lattice statistics problems in general. The approach is based on the following observation: Instead of calculating the quantities of interest, e.g., $c_N$, directly, it is often easier to calculate the *function*, e.g., $\chi(z)$ in (2.3), which generates these quantities. Once the generating function is known, various characteristic quantities are easily obtained by, e.g., taking the appropriate derivatives. In order to illustrate the technique, let us consider the fully directed SAW on the two-dimensional square lattice, as illustrated in Fig. 1 (d).

The fully directed SAW (FDSAW) on the square lattice consists of steps which start at the origin and are allowed to proceed only in the $+X$ or $+Y$ directions, see Fig. 1 (d). Here we take the $X$ and $Y$ axes to coincide with principal directions on the square lattice. The preferred axis of directedness of the walk ("time axis") is the $Y = X$ line: each $+a\hat{X}$ and $+a\hat{Y}$ step advances the walk by $a/\sqrt{2}$ along the time axis, with $a$ being the lattice spacing. We will carry out the calculation in the fixed fugacity, $z$, grand-canonical type ensemble. For each $N$-step SAW we assign statistical weight $z^N$. To make the model slightly more general we also introduce statistical weight $w$ for every turn (of 90°) in the walk. (These turn-weighted models are used to describe single-chain rod-to-coil transition which we study in the next section.) Note also that ordinary FDSAW is obtained by setting $w = 1$. The partition function for the $N$-step $T$-turn walk is then

$$Z(z; w) = \sum_{\text{all walks}} z^N w^T. \tag{2.10}$$

Throughout this section we will not include walks with $N = 0$ steps, i.e., $c_0 \equiv 0$ in sums like (2.10). Thus, the susceptibility $\chi(z)$, defined by (2.3), is obtained from (2.10) as $\chi(z) = 1 + Z(z; 1)$. To calculate the appropriate generating function we assign statistical weights: $x$ per each $+a\hat{X}$ step and $y$ per each $+a\hat{Y}$ step. Let $n_x$ and $n_y$ denote the number of $+X$ and $+Y$ steps in a given walk, so that $N = n_x + n_y$. It is convenient to introduce three generating functions,

$$G(x, y; w) \equiv \sum_{\text{all walks}} x^{n_x} y^{n_y} w^T = G_x + G_y, \tag{2.11}$$

$$G_x(x, y; w) \equiv \sum^{(x)} x^{n_x} y^{n_y} w^T, \tag{2.12}$$

$$G_y(x, y; w) \equiv \sum^{(y)} x^{n_x} y^{n_y} w^T. \tag{2.13}$$

The sum in (2.11) is over all possible walks. The sum in (2.12) is over all walks that start with the $+a\hat{X}$ step, while the sum in (2.13) is over walks that start with the $+a\hat{Y}$ step. We are actually interested in the total generating function $G(x, y; w)$ since $Z(z; w) \equiv G(z, z; w)$. However, the restricted generating functions $G_x$ and $G_y$ are easier to calculate. In particular, they satisfy the recursion relations which are illustrated schematically in Fig. 2, and which take the form

$$G_x = x + xG_x + wxG_y, \tag{2.14}$$

$$G_y = y + yG_y + wyG_x. \tag{2.15}$$

Solving for $G_x$ and $G_y$ and using (2.11), we get

$$G(x, y; w) = \frac{x + y + 2(w - 1)xy}{(1 - x)(1 - y) - w^2 xy}. \tag{2.16}$$

Thus, the required partition function is

$$Z(z; w) = G(z, z; w) = \frac{2z}{1 - (w + 1)z}. \tag{2.17}$$

Note that $Z(z; w)$ has a simple pole singularity at $z_c = 1/(1 + w)$. With $w = 1$, we obtain by expanding (2.17) that the number of ordinary FDSAW of length $N$ is

$$c_N = 2^N, \tag{2.18}$$

giving $\gamma = 1$ and $\mu = 2$. (This last result can be obtained directly if we recall that the probability for the FDSAW on the square lattice to proceed to any allowed lattice point equals $1/2$, and that the steps are statistically independent.)

The exponents $\nu_\parallel$ and $\nu_\perp$ can be also obtained from the generating function. Specifically, we define the $k$th-moment parallel correlation length

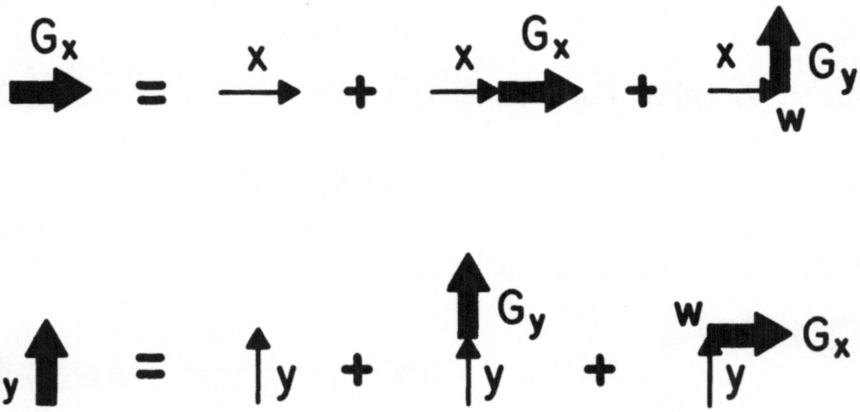

**Figure 2**

Diagrammatic representation of the recursion relations (2.14) and (2.15). Full
arrows denote all possible walks with the initial step along that arrow.

$$[\xi_{\parallel}^{(k)}(z;w)]^k \equiv \frac{\sum r_{\parallel}^k z^N w^T}{Z(z;w)}, \tag{2.19}$$

where $r_{\parallel}$ is the displacement parallel to the directed axis, of a given $N$-step $T$-turn walk. For the FDSAW this quantity is in fact given by $(a/\sqrt{2})(n_x + n_y)$. Using (2.11) and (2.19), the $k$th-moment parallel correlation length is then represented as

$$[\xi_{\parallel}^{(k)}(z;w)]^k = Z^{-1} \left(\frac{a}{\sqrt{2}}\right)^k \left(z\frac{\partial}{\partial z}\right)^k G(z,z;w). \tag{2.20}$$

Similarly, we define the $k$th-moment perpendicular correlation length

$$[\xi_{\perp}^{(k)}(z;w)]^k \equiv \frac{\sum r_{\perp}^k z^N w^T}{Z(z;w)}, \tag{2.21}$$

where $r_{\perp}$ is the displacement transverse to the directed axis, of a given $N$-step $T$-turn walk. This displacement can have both positive and negative values (symmetrically distributed), and is given by $(a/\sqrt{2})(n_x - n_y)$. In order to express the $k$th-moment perpendicular correlation length (2.21) in terms of the derivative of the generating function, we put $x = pz$ and $y = p^{-1}z$ in (2.11). Then (2.21) yields

$$[\xi_{\perp}^{(k)}(z;w)]^k = Z^{-1} \left(\frac{a}{\sqrt{2}}\right)^k \left[\left(p\frac{\partial}{\partial p}\right)^k G(pz,p^{-1}z;w)\right]_{p=1}. \tag{2.22}$$

Note that $\xi_{\perp}^{(k)}$ vanishes for all odd values of $k$. Using the result (2.16), and calculating the derivatives as indicated in (2.20) and (2.22), we get

$$\xi_{\parallel}^{(1)}(z;w) = \left(\frac{a}{\sqrt{2}}\right) \frac{1}{1 - (1+w)z}, \tag{2.23}$$

and

$$[\xi_\perp^{(2)}(z;w)]^2 = \left(\frac{a^2}{2}\right)\frac{1-(w-1)z}{[1+(w-1)z][1-(w+1)z]}. \tag{2.24}$$

Thus, we obtain $\nu_\parallel = 1$ and $\nu_\perp = 1/2$, as expected for DSAW (see discussion following (2.9)). These results show that DSAW models have no really long-range self-avoidance effects.

The above derivation for planar walks can be easily generalized to higher dimensional FDSAW on hypercubic lattices. Specifically, in analogy with (2.11), for the $d$-dimensional FDSAW we introduce the generating function

$$G(x_1, x_2, ..., x_d; w) = \sum_{i=1}^{d} G_i(x_1, ..., x_d; w), \tag{2.25}$$

where the restricted generating functions $G_i$ correspond to walks which start with $+a\hat{X}_i$ steps ($i = 1, ..., d$). The restricted functions satisfy a $d$-dimensional generalization of the recursion relations (2.14) and (2.15), i.e.,

$$G_i = x_i + x_i G_i + w x_i \sum_{j\neq i} G_j, \tag{2.26}$$

which express the fact that a walk starting with the $+a\hat{X}_i$ step can either end (weight $x_i$), or continue along $+X_i$ (weight $x_i G_i$), or turn to $+X_j$ (weights $w x_i G_j$). Straightforward calculations then yield

$$G(x_1, ..., x_d; w) = \frac{S_d}{1 - w S_d}, \quad \text{with} \quad S_d \equiv \sum_{i=1}^{d} \frac{x_i}{1-(1-w)x_i}. \tag{2.27}$$

Thus, the partition function,

$$Z(z; w) \equiv G(z, ..., z; w) = \frac{zd}{1-[1+(d-1)w]z}, \tag{2.28}$$

for FDSAW is obtained. [Note that for $d = 2$ in (2.28), we reproduce (2.17).] It has a simple pole singularity at $z_c = 1/[1 + (d - 1)w]$. Expanding (2.28) with $w = 1$ we obtain $c_N = d^N$, i.e., by comparing with (2.1), $\gamma = 1$ and $\mu = d$ for the ordinary $d$-dimensional FDSAW.

The exponents $\nu_\parallel$ and $\nu_\perp$ characterizing the sizes of this $d$-dimensional walk can be obtained by generalizing the procedure used in the $d = 2$ case. In particular, it is easy to see that the parallel displacement $r_\parallel$ after $N$ steps is

$$r_\parallel = \frac{a}{\sqrt{d}} N, \tag{2.29}$$

for a $d$-dimensional walk. Therefore, using definition (2.19), the first moment parallel correlation length is

$$\xi_\parallel^{(1)}(z; w) = \frac{1}{Z(z; w)} \left(\frac{a}{\sqrt{d}}\right) z \frac{\partial}{\partial z} Z(z; w). \tag{2.30}$$

Substituting the result (2.28) in this equation we get

$$\xi_\parallel^{(1)} = \left(\frac{a}{\sqrt{d}}\right) \frac{1}{1 - [1 + (d - 1)w]z}, \tag{2.31}$$

from which $\nu_\parallel = 1$ is obtained. Higher moments can be similarly calculated. For example, by (2.19) and (2.29) we have

$$[\xi_\parallel^{(2)}(z; w)]^2 = \frac{a^2}{d} \frac{1}{Z(z; w)} \left(z \frac{\partial}{\partial z}\right)^2 Z(z; w), \tag{2.32}$$

which yields

$$[\xi_\parallel^{(2)}(z; w)]^2 = \frac{a^2}{d} \frac{1 + [1 + (d - 1)w]z}{\{1 - [1 + (d - 1)w]z\}^2}, \tag{2.33}$$

where we used (2.28).

The first-moment perpendicular correlation length vanishes by symmetry. The second moment is defined by (2.21), and can be calculated by noting that

$$r_\perp^2 = r^2 - r_\parallel^2 = a^2 \sum_{i=1}^{d} n_i^2 - r_\parallel^2, \tag{2.34}$$

from which

$$[\xi_\parallel^{(2)}]^2 + [\xi_\perp^{(2)}]^2 = \frac{a^2}{Z(z;w)} \sum_{i=1}^{d} \left[\left(x_i \frac{\partial}{\partial x_i}\right)^2 G(x_1, ..., x_d; w)\right]_{x_j = z}. \tag{2.35}$$

The resulting expression for $\xi_\perp^{(2)}$ is

$$[\xi_\perp^{(2)}(z;w)]^2 = \frac{a^2}{d} \frac{(d-1)[1 + (1-w)z]}{[1-(1-w)z]\{1 - z[1 + (d-1)w]\}}, \tag{2.36}$$

where we used (2.28) and (2.33). Therefore, we obtain $\nu_\perp = 1/2$ for FDSAW on a $d$-dimensional lattice. Clearly, the correlation length exponents $\nu_\parallel = 1$ and $\nu_\perp = 1/2$ are independent of dimensionality.

The full directedness of the SAW can be relaxed somewhat if, for instance, in the $d = 2$ case, steps in both $+Y$ and $-Y$ direction are allowed. One such walk is shown in Fig. 1 (c). The appropriate $d$-dimensional generalization is obtained when positive and negative steps along *one* of the $d$ directions, say $\pm a\hat{X}_d$, are allowed. With these restrictions a *partially directed* SAW (PDSAW) is obtained. Note that for such walks the self-avoidance condition reduces to "no immediate return" restriction. We now discuss the general $d$-dimensional PDSAW, from which the planar case is obtained simply by setting $d = 2$.

Let $n_i$ $(i = 1, ..., d-1)$ and $n_\pm$ denote the number of the $+a\hat{X}_1, ..., +a\hat{X}_{d-1}$, and $\pm a\hat{X}_d$ steps in a given $N$-step walk, with

$$N = \sum_{i=1}^{d-1} n_i + n_+ + n_-. \tag{2.37}$$

The generating function for the $N$-step, $T$-turn, $d$-dimensional walk is defined by

$$G(x_1, ..., x_{d-1}; x_+, x_-; w) = \sum_{\text{all walks}} \left( w^T x_+^{n_+} x_-^{n_-} \prod_{i=1}^{d-1} x_i^{n_i} \right), \qquad (2.38)$$

where the notation for the step weights is self-explanatory. The partial generating functions for walks with the first steps $+a\hat{X}_1, ..., +a\hat{X}_{d-1}, \pm a\hat{X}_d$ will be denoted $G_1, ..., G_{d-1}, G_\pm$, respectively. They satisfy the recursion relations

$$G_i = x_i + x_i G_i + w x_i \left( \sum_{j=1, j \neq i}^{d-1} G_j + G_+ + G_- \right), \qquad (2.39)$$

for $i < d$. For $G_\pm$ we have

$$G_\pm = x_\pm + x_\pm G_\pm + w x_\pm F \quad \text{with} \quad F = \sum_{i=1}^{d-1} G_i. \qquad (2.40)$$

Summing the equations (2.39) over $i = 1, ..., d-1$ and combining the result with (2.40) we obtain a system of three linear equations for $F$ and $G_\pm$. These are easily solved to yield the final result

$$G(x_1, ..., x_{d-1}; x_+, x_-; w) = \frac{S_{d-1} + T + wTS_{d-1}}{1 - wS_{d-1} - w^2 T S_{d-1}}, \qquad (2.41)$$

with $S_d$ defined as in (2.27) and

$$T(x_\pm) \equiv \frac{x_+}{1 - x_+} + \frac{x_-}{1 - x_-}. \qquad (2.42)$$

Note that the relation (2.27) for FDSAW is obtained from (2.41) if either $x_+$ or $x_-$ is set equal to zero. The partition function $Z(z; w) \equiv G(z, ..., z; z, z; w)$ for PDSAW is

$$Z(z; w) = \frac{z[(d+1)(1-z) + 2dwz]}{1 - 2z - (d-2)wz - [2(d-1)w^2 - (d-2)w - 1]z^2}, \qquad (2.43)$$

with the simple pole singularity at

$$z_c(w) = \frac{2}{2 + w[d - 2 + \sqrt{d^2 + 4d - 4}]}. \tag{2.44}$$

Since (2.43) is of the form $\sim (z_c - z)^{-1}$, we obtain that the number, $c_N$, of PDSAW of length $N$ is given by (2.1) with $\gamma = 1$ and $\mu = 1/z_c(w = 1; d)$, for all values of $d$.

The correlation lengths can also be calculated from the generating function (2.41). Note that the parallel displacement is given by

$$r_\| = \frac{a}{(d-1)^{1/2}}(n_1 + ... + n_{d-1}), \tag{2.45}$$

so that the $k$th-moment correlation length, defined by (2.19), can be expressed as

$$[\xi_\|^{(k)}]^k = Z^{-1} \left(\frac{a}{\sqrt{d-1}}\right)^k \left[\left(z\frac{\partial}{\partial z}\right)^k G(z, ..., z; x_\pm; w)\right]_{x_\pm = z}. \tag{2.46}$$

Using the results (2.41) and (2.46), with $k = 1$, we obtain after a long but straightforward algebra that $\xi_\|^{(1)} \sim 1/(z_c - z)$ (explicit expression is too long to be reproduced here). Therefore $\nu_\| = 1$ for PDSAW in any dimension $d$, as anticipated. Similarly, by using the definition (2.21), and relations (2.34) with (2.45), and (2.46) with $k = 2$, we can also obtain $\xi_\perp^{(2)}$. After a long algebra (not given here) we find $[\xi_\perp^{(2)}]^2 \sim 1/(z_c - z)$, and the result $\nu_\perp = 1/2$ finally follows.

In summary, we have established by exact calculation that $\gamma = 1$, $\nu_\| = 1$, and $\nu_\perp = 1/2$ for two types of DSAW, independent of the type and dimensionality of the lattice. In fact, exact calculations for other $2d$ lattices (e.g., triangular) have been performed. Such calculations confirm that the exponent values do not depend on the lattice type.

## C. Rod-to-coil transition of linear polymers

In this section we use the results derived in Section B to analyze the rod-to-coil transition of polymer chains. We will particularly concentrate on the scaling properties of this transition. Rod-to-coil transitions in general exhibit an interesting property of scaling function nonuniversality which, in the case of directed models, can be demonstrated explicitly. In the limit of infinite dimensionality, however, the universality is restored.

To properly model the physical situation characterizing the rod-to-coil transition, we consider a "stiff chain" limit of a small statistical weight $w$ for every turn in the walk. As already discussed, the appropriate partition function in the fixed fugacity ensemble of such turn-weighted DSAW model is

$$Z(z; w) = \sum_{\text{all walks}} z^N w^T, \tag{2.47}$$

for an $N$-step $T$-turn walk. As discussed in Section B, the walk with $N = 0$ steps is not included in this sum, i.e., $c_0 \equiv 0$ here. The physics of the rod-to-coil transition is already apparent in (2.47): For small enough values of $w$, only the walks with no turns or small number of turns significantly contribute to the partition function, i.e., the fully extended, rod-like configurations statistically dominate. On the other hand, when $w$ is not too small, the walks with many turns dominate in (2.47), and the coiled configurations are favored. The crossover between the two behaviors is termed the rod-to-coil transition.

Let us now consider scaling in the rod-to-coil transition regime. This regime is defined by $w \to 0$, $N \to \infty$, with the scaling combination

$$\tilde{\omega} \equiv wN, \tag{2.48}$$

taking values $O(1)$. That $\tilde{\omega}$ is linear in both $w$ and $N$ is a nontrivial property implied by exact results for the Gaussian and directed walks, and by numerical studies for isotropic SAWs. In the fixed fugacity ensemble, we will consider the equivalent of the scaling combination (2.48), but with the *average* number of steps, $N(z; w)$, where

$$N(z; w) = \frac{\sum N \, z^N \, w^T}{\sum z^N \, w^T} \equiv z \frac{\partial \ln Z(z; w)}{\partial z}, \tag{2.49}$$

instead of $N$. The summation in (2.49) is over "all walks", as in (2.47). For *fixed* $w$, the quantity $N(z; w)$ has a simple pole singularity at $z_c(w)$. Since $Z(z; w)$ has a power law singularity with the critical exponent $\gamma$, one can further conclude that in the limit $z \to z_c(w)$, with $w$ fixed,

$$N(z; w) \approx \frac{\gamma z_c(w)}{z_c(w) - z}. \tag{2.50}$$

Recall that for DSAW, $\gamma \equiv 1$. Relation (2.50) takes a particularly simple form for $w = 0$ since $z_c(0) = 1$. We have

$$N(z; 0) \equiv \bar{N}(z) \equiv \frac{1}{1 - z}. \tag{2.51}$$

The scaling combination analogous to (2.48) can be defined as $wN(z; w)$ or $wN(z; 0) = \omega \bar{N}$, and the two choices are equivalent up to a redefinition of the scaling function, see below. We use the second combination

$$\omega \equiv wN(z; 0) \equiv w\bar{N} \equiv w/(1 - z) \equiv \omega(z; w), \tag{2.52}$$

which in the scaling limit ($w \to 0$, $\bar{N} \to \infty$), takes values $O(1)$. Physically, this means that for fixed $w$, long chains will be coiled provided $N >> w^{-1}$. For fixed length $N$, stiff chains ($w << N^{-1}$) will be rodlike. Thus, the transition occurs when $w$ and $N^{-1}$ are comparable, i.e., $\omega$ is of order one. The use of $\bar{N}$ is for technical convenience only; in principle the scaling relations could be formulated in terms of $N(z; w)$.

The scaling relations take the form

$$\frac{Z(z; w)}{Z(z; 0)} \equiv \frac{1}{1 - (d-1)w(\bar{N} - 1)} \approx \frac{1}{1 - (d-1)\omega} = A(c\omega), \qquad (2.53)$$

and

$$\frac{N(z; w)}{N(z; 0)} \approx \frac{1}{1 - (d-1)\omega} = B(c\omega), \qquad (2.54)$$

where we used (2.28) for the FDSAW, and (2.49) to obtain $N(z; w)$. The *scaling functions* $A$ and $B$ are defined in terms of the "linear scaling field"

$$g \equiv c\omega, \qquad (2.55)$$

where the metric factor $c$ can be determined from the fixed-$w$ critical point value $z_c(w)$. Indeed, for fixed $w > 0$, the functions $Z(z; w)$ and $N(z; w)$ diverge as $[z - z_c(w)]^{-1}$ in the limit $z \to z_c(w)$. The metric factor is given by the scaling limiting behavior of the combination

$$[1 - z_c(w)]\bar{N}(z) \approx c\omega, \qquad (2.56)$$

which yields

$$c_{FD} = d - 1. \qquad (2.57)$$

This fixes the scaling functions,

$$A_{FD}(g) = B_{FD}(g) = (1 - g)^{-1}. \qquad (2.58)$$

Similar results can be derived in the case of $d$-dimensional PDSAW. Using the result (2.43) and the scaling function as in (2.53), we get

$$A_{PD}(c\omega) = \frac{1 + d(1 + 2\omega)}{(d+1)[1 - (d-2)\omega - 2(d-1)\omega^2]}. \tag{2.59}$$

The metric factor can be identified by using the location of the singularity of $Z(z;w)$. It is given by (2.44) and, when combined with (2.49) and (2.56), gives

$$c_{PD} = \frac{d - 2 + \sqrt{d^2 + 4d - 4}}{2}. \tag{2.60}$$

The expression for the average number of steps $N(z;w)$ is very long. (It can be calculated straightforwardly by the use of (2.49) and (2.43).) We only report here

$$N(z;0) = (1 - z)^{-1} \equiv \bar{N}(z). \tag{2.61}$$

The appropriate scaling function, defined as in (2.54), is then

$$B_{PD}(c\omega) = \frac{1 + d(1 + 4\omega) + 2(2d - 1)\omega^2}{[1 + d(1 + 2\omega)][1 - (d-2)\omega - 2(d-1)\omega^2]}. \tag{2.62}$$

In order to derive scaling relations for the correlation lengths we form ratios

$$\frac{\xi_{\parallel}^{(k)}(z;w)}{\xi_{\parallel}^{(k)}(z;0)} \approx P^{(k)}(c\omega), \qquad k \geq 1, \tag{2.63}$$

and

$$\frac{\xi_{\perp}^{(k)}(z;w)}{\xi_{\perp}^{(k)}(z;0)} \approx Q^{(k)}(c\omega), \qquad k \text{ even}. \tag{2.64}$$

The scaling functions $P^{(k)}(c\omega)$ and $Q^{(k)}(c\omega)$ can be evaluated analytically by the use of the results already derived for the $d$-dimensional FDSAW and PDSAW. Explicit expressions for the $k = 1, 2$ correlation length scaling functions for FDSAW are

$$P_{FD}^{(1)}(c\omega) = P_{FD}^{(2)}(c\omega) = \frac{1}{1-(d-1)\omega}, \tag{2.65}$$

and

$$Q_{FD}^{(2)}(c\omega) = \frac{1}{\sqrt{(1+\omega)[1-(d-1)\omega]}}, \tag{2.66}$$

where we used (2.31), (2.33), and (2.36). Similarly, for $d$-dimensional PDSAW, the scaling functions can be obtained by the use of (2.43) and (2.46). We get

$$P_{PD}^{(1)}(c\omega) = \frac{(d+1)(1+2\omega)^2}{[1+d(1+2\omega)][1-(d-2)\omega-2(d-1)\omega^2]}, \tag{2.67}$$

$$P_{PD}^{(2)}(c\omega) = \left[\frac{d+1}{1+d(1+2\omega)}\right]^{1/2} \frac{1+2\omega}{1-(d-2)\omega-2(d-1)\omega^2}, \tag{2.68}$$

and

$$[Q_{PD}^{(2)}(c\omega)]^2 = \frac{(d+1)[d+2(2d-1)\omega(1+\omega)+2\omega^3]}{d(1+\omega)[1+d(1+2\omega)][1-(d-2)\omega-2(d-1)\omega^2]}. \tag{2.69}$$

It is obvious from these results that contrary to the conventional "scaling" wisdom, the scaling functions are nonuniversal (i.e., model dependent) for all finite dimensionalities $d$. However, it turns out that the FDSAW and PDSAW have the same scaling properties in the infinite dimensionality limit. Note that the scaling-field combination

$$g = c\omega, \tag{2.70}$$

contains, in the $d \to \infty$ limit, additional unbounded parameter since for both models considered $c \approx d$, as $d \to \infty$. Thus, as long as we define the scaling limit with fixed $dw\bar{N} = O(1)$, with $d$, $w^{-1}$, and $\bar{N}$ large, we can formally expand the scaling functions in powers of $1/d$. We thus substitute $g/c(d)$ for $\omega$ and regard the scaling functions as functions of $g$ and $1/d$. The leading order result (in the $1/d$ expansion) is surprisingly simple,

$$A(g;d), \quad B(g;d), \quad P^{(1,2)}(g;d), \quad [Q^{(2)}(g;d)]^2 \approx (1-g)^{-1}, \qquad (2.71)$$

for both models. Thus, in the limit of infinite dimensionality, the universality of the scaling functions is restored. Finally, we note that the nonuniversality of the type just described can be also obtained for other types of lattices, and also for isotropic SAWs. It is related to the nonexistence of the field-theoretical scaling-limiting continuum description of the stiff polymer chains.

## D. Finite-size scaling results

Until now we have considered scaling properties of polymers on infinite lattices. However, in all numerical simulations and in some experiments, polymers are confined to lattices (pores) of finite size. Experimentally, it is thus important to understand the influence of finiteness of the pore on the scaling properties and obtain corrections due to the finite size. Once the finite-size effects are known, the thermodynamic limit (infinite-size) quantities can be obtained by proper extrapolation of the numerical data.

Consider a PDSAW on a finite square lattice of $L_{\parallel} \times L_{\perp}$ sites along the $X$ and $Y$ directions, respectively. [The walk starts from some origin ($X = 0, Y = 0$) and can advance only in $+X$ and $\pm Y$ directions.] For simplicity we consider periodic boundary conditions in the $Y$ direction, so that the walks are actually on the cylinder of length $L_{\parallel}$ and circumference $L_{\perp}$. (Here and further below we assume the lattice constant $a$ to be unity.) As in the case of infinite lattice, we analyze the properties of this walk by the generating function technique.

To specify the walk we assign weights $x$, $y_+$, and $y_-$ to every horizontal $+\hat{X}$, upward $+\hat{Y}$, and downward $-\hat{Y}$ step. We also denote by $C_{N n_+ n_-}(L_{\|}, L_\perp)$ the number of $N$-step walks which start at the origin $(X = 0, Y = 0)$ and have $n_+$ upward and $n_-$ downward steps.

Consider first, for simplicity, the case $L_{\|} = \infty$. The generating function is

$$G(x, y_+, y_-; \infty, L_\perp) \equiv \sum_{\text{all walks}} C_{N n_+ n_-}(\infty, L_\perp) x^{N - n_+ - n_-} y_+^{n_+} y_-^{n_-}, \qquad (2.72)$$

where the notation is self-explanatory. The sum over "all walks" now includes also the walks of $N = 0$ steps, i.e., we take $c_0 = 1$ for convenience, in this section. This sum can be further expressed as

$$G(x, y_+, y_-; \infty, L_\perp) = H[1 + xH + (xH)^2 + ...] = \frac{H}{1 - xH}, \qquad (2.73)$$

where $H$ is the generating function for a walk parallel to the $Y$-axis, i.e.,

$$H(y_+, y_-; L_\perp) = 1 + (y_+ + y_+^2 + ... + y_+^{L_\perp - 1}) + (y_- + y_-^2 + ... + y_-^{L_\perp - 1})$$
$$= 1 + \frac{y_+ - y_+^{L_\perp}}{1 - y_+} + \frac{y_- - y_-^{L_\perp}}{1 - y_-}. \qquad (2.74)$$

The total number, $c_N$, of $N$-step walks is obtained from the susceptibility

$$\chi(z; \infty, L_\perp) \equiv G(z, z, z; \infty, L_\perp) = \sum_{N=0}^{\infty} c_N(\infty, L_\perp) z^N. \qquad (2.75)$$

We get

$$\chi(z;\infty,L_\perp) = \frac{1+z-2z^{L_\perp}}{1-2z-z^2+2z^{L_\perp+1}}. \tag{2.76}$$

For $L_\perp = \infty$ this yields the result (2.43), with $d = 2$, $w = 1$, and $\chi(z;\infty,\infty) = 1+$ $Z(z;1)$. The susceptibility has a simple pole singularity at the $z$-value $z_c = \sqrt{2}-1$, (compare (2.44) with $d = 2$, $w = 1$), which represents a bulk critical point near which $\chi(z;\infty;\infty) \sim (z_c-z)^{-\gamma}$, with $\gamma = 1$. Parallel and perpendicular correlation lengths are defined by the relations analogous to (2.4). Specifically,

$$\xi_\parallel(z;\infty,L_\perp) \equiv \frac{\sum c_N \langle R_{N\parallel}\rangle z^N}{\sum c_N z^N}, \tag{2.77}$$

with the mean parallel displacement defined by

$$\langle R_{N\parallel}(\infty,L_\perp)\rangle \equiv c_N^{-1} \sum_{n_+,n_-} (N - n_+ - n_-)C_{Nn_+n_-}. \tag{2.78}$$

Note that (2.77), etc., differ from the quantities defined in Section B (i.e., $\xi_\parallel^{(1)}/a$) by the choice of $c_0 = 1$ here. As mentioned, this does not affect the behavior near $z_c$. From (2.72), (2.77), and (2.78) we have

$$\xi_\parallel(z;\infty,L_\perp) = z\left(\frac{\partial}{\partial x}\ln G\right)_{x=y_+=y_-=z} = z\chi(z;\infty,L_\perp). \tag{2.79}$$

Defining

$$t \equiv (z_c - z)/z_c, \tag{2.80}$$

one rediscovers the bulk singular behavior with $\nu_\parallel \equiv 1$,

$$\xi_\parallel(z;\infty,\infty) \approx \frac{1}{2}t^{-\nu_\parallel}. \tag{2.81}$$

Similarly, for the perpendicular correlation length, we use

$$\xi_\perp^2(z; \infty, L_\perp) \equiv \frac{\sum c_N \langle R_{N\perp}^2 \rangle z^N}{\sum c_N z^N}, \tag{2.82}$$

with

$$\langle R_{N\perp}^2 \rangle \equiv c_N^{-1} \sum_{n_+, n_-} (n_+ - n_-)^2 C_{N n_+ n_-}. \tag{2.83}$$

This yields

$$\xi_\perp^2(z; \infty, L_\perp) = \hat{D}_\perp G(x, y_+, y_-; \infty, L_\perp), \tag{2.84}$$

where $\quad \hat{D}_\perp f \equiv (1/f) \left[ z^2 \left( \frac{\partial}{\partial y_+} - \frac{\partial}{\partial y_-} \right)^2 f + 2z \frac{\partial f}{\partial y_+} \right]_{x=y_+=y_-=z}$

Here we adopt a convention that a walk with total of, say, $2L_\perp - 1$ upward steps and no downward steps is regarded as having net vertical displacement of $2L_\perp - 1$ from the origin, *not* 1 as measured from the shortest path back to the origin. This is mathematically convenient and in the limit $L_\perp \to \infty$ reduces to the usual bulk definition of vertical displacement. We thus have

$$\xi_\perp^2(z; \infty, L_\perp) = [\xi_\parallel(z; \infty, L_\perp) + 1]\hat{D}_\perp H(y_+, y_-; L_\perp), \tag{2.85}$$

with

$$\hat{D}_\perp H(y_+, y_-; L_\perp) = \frac{2z}{(1-z)^2} + O(L_\perp^2 z^{L_\perp}). \tag{2.86}$$

In the bulk limit this reduces to

$$\xi_\perp^2(z; \infty; \infty) \approx \frac{1}{2z_c} t^{-2\nu_\perp}, \quad \text{with} \quad \nu_\perp = \frac{1}{2}, \tag{2.87}$$

as expected.

Let us now examine the correlation lengths (2.79) and (2.85) for finite $L_\perp$. Note that (2.76) implies that the correlation lengths diverge (with exponents $\nu_\parallel = 1$ and $\nu_\perp = 1/2$), but that the critical point is shifted to the $z$-value

$$z_0(L_\perp) = z_c[1 + \frac{z_c^{L_\perp}}{\sqrt{2}} + O(L_\perp z_c^{2L_\perp})]. \tag{2.88}$$

Thus, *at $z = z_c$, $\xi_\parallel$ diverges exponentially with $L_\perp$:*

$$\xi_\parallel(z_c; \infty, L_\perp) \approx \frac{1}{\sqrt{2}} z_c^{-L_\perp}. \tag{2.89}$$

This implies that in the limit of bulk criticality ($z \to z_c^-, L_\perp \to \infty$) the new length scale appears in addition to the usual bulk correlation length $\xi_\parallel(z; \infty, \infty)$. In order to present this in a more conventional form, we define $\sigma \equiv -\ln z_c > 0$ and introduce an exponentially divergent length

$$l_\parallel(L_\perp) \equiv e^{\sigma L_\perp} = z_c^{-L_\perp} \sim \xi_\parallel(z_c; \infty, L_\perp). \tag{2.90}$$

Using (2.79) and (2.86) we can express $\xi_\parallel$ in the scaling form

$$\xi_\parallel(z; \infty, L_\perp) \approx \frac{1}{2t} \mathcal{W}[2t l_\parallel(L_\perp)], \tag{2.91}$$

with the *scaling function* given by

$$\mathcal{W}(v) \equiv \frac{v}{\sqrt{2} + v}. \tag{2.92}$$

With (2.80), this can be finally written in the form

$$\xi_\parallel(z; \infty, L_\perp) \approx \xi_\parallel(z; \infty, \infty) \mathcal{W}\left( \frac{l_\parallel(L_\perp)}{\xi_\parallel(z; \infty, \infty)} \right). \tag{2.93}$$

This equation shows that $\xi_\parallel$ of the finite system has a scaling form consistent with the *weak scaling hypothesis*, i.e., it involves the additional length $l_\parallel(L_\perp)$ which grows exponentially with $L_\perp$. For the correlation length to obey the strong scaling hypothesis, with $L_\parallel = \infty$, one would have

$$\xi_i(t; \infty, L_\perp) \approx L_\perp^{\nu_i/\nu_\perp} \mathcal{Y}(L_\perp t^{\nu_\perp}), \quad i = \parallel, \perp . \tag{2.94}$$

Instead, we have

$$\xi_\parallel(z; \infty, L_\perp) \approx e^{\sigma L_\perp} \bar{\mathcal{Y}}(e^{\sigma L_\perp} t^{\nu_\parallel}), \tag{2.95}$$

with a similar result for the $\xi_\perp$, see below. This is reminiscent of the scaling properties near a first-order transition in, e.g., Ising-type models (systems with scalar order parameter) on cylinders, where the corresponding length, $\xi_1(L_\perp) \sim \exp(\tau L_\perp^{d-1})$, describes the average separation of interfaces orthogonal to the cylinder axis, with $\tau$ being the surface tension, and $L_\perp^{d-1}$ the cross-section of the cylinder. The domains of single phase are of the average length $\xi_1$ which diverges exponentially in the limit $L_\perp \to \infty$.

In the case of finite $L_\parallel$, the results are similar. The generating function is now

$$G(x, y_+, y_-; L_\parallel, L_\perp) = H[1 + xH + (xH)^2 + ... + (xH)^{L_\parallel - 1}] = \frac{H[1 - (xH)^{L_\parallel}]}{1 - xH}, \tag{2.96}$$

so that $\xi_\parallel$, defined by (2.79), becomes

$$\xi_\parallel(z; L_\parallel, L_\perp) = \xi_\parallel(z; \infty, L_\perp) - \frac{L_\parallel(zH_0)^{L_\parallel}}{1 - (zH_0)^{L_\parallel}}, \tag{2.97}$$

with

$$H_0 = H(z, z; L_\perp) = \frac{1 + z - 2z^{L_\perp}}{1 - z}.$$
(2.98)

It has the asymptotic scaling form near bulk criticality

$$\xi_\parallel(z; L_\parallel, L_\perp) \approx \frac{1}{2t}\bar{\mathcal{W}}(2tL_\parallel, 2tl_\parallel),$$
(2.99)

with the scaling function

$$\bar{\mathcal{W}}(u, v) \equiv \mathcal{W}(v) - \frac{u}{e^{u/\mathcal{W}(v)} - 1},$$
(2.100)

which has again the form of the weak scaling hypothesis. The perpendicular correlation length is still given by (2.85), with $L_\parallel < \infty$, and in the scaling region takes the form

$$\xi_\perp^2(z; L_\perp, L_\parallel) \approx \frac{1}{z_c}\xi_\parallel(z; L_\parallel, L_\perp),$$
(2.101)

which implies that its scaling properties are similar to those of $\xi_\parallel$.

In summary, the characteristic finite-size properties of the DSAWs are a consequence of their extreme anisotropy. Note, however, that scaling with exponential length scales has not been observed in the case of directed percolation, for instance. Thus, the finite-size scaling properties of directed walk models have interesting features which show that in models with continuous bulk transitions, but with anisotropic divergence of correlations, the conventional "power-law" scaling forms may not always hold.

## E. Selected literature

The following list includes several books which can provide a general exposition on the topics of random walks and polymer scaling.

**K.F. Freed** (1987) *Renormalization Group Theory of Macromolecules*
(Wiley, New York)

**P.-G. de Gennes** (1979) *Scaling Concepts in Polymer Physics* (Cornell, Ithaca)

**H. Yamakawa** (1971) *Modern Theory of Polymer Solutions*
(Harper & Row, New York)

**P.J. Flory** (1969) *Statistical Mechanics of Chain Molecules* (Wiley, New York)

**M.V. Volkenstein** (1963) *Configurational Statistics of Polymeric Chains*
(Wiley, New York)

**P.J. Flory** (1953) *Principles of Polymer Chemistry* (Cornell, Ithaca)

**W. Feller** (1951) *An Introduction to Probability Theory and Its Applications*
(Wiley, New York)

The results for directed walks, described in this chapter, have been collected from the following original articles: V. Privman and N.M. Švrakić, J. Statist. Phys. **50**, 81 (1988); V. Privman and H.L. Frisch, J. Chem. Phys. **88**, 469 (1988); A.M. Szpilka and V. Privman, Phys. Rev. B**28**, 6613 (1983).

# III. LINE INTERFACES IN TWO DIMENSIONS: SOLID-ON-SOLID MODELS

Solid–on–Solid (SOS) models have been extensively used to describe the properties of interfaces separating coexisting thermodynamic phases. In this chapter we study such models in $2d$, i.e., for line interfaces. Specifically, we concentrate on the transfer matrix calculations appropriate for bound and fluctuating interfaces and analyze scaling and finite-size properties of the SOS models. In Section A, definitions of the unrestricted and restricted lattice SOS models are given and basic features of the wetting transition in the presence of "contact" short-range substrate forces are presented. In Sections B and C we consider wetting transitions in the presence of short- and long-range interactions and describe various results obtained in the continuous limit and by other techniques, including the generating function and continued fraction methods. Finite-size scaling for wetting (i.e., interfacial unbinding) is studied in Section D, while Section E describes finite-size results for fluctuating (unbound) interfaces. Finally, Section F contains the list of selected literature.

## A. Wetting transitions in the SOS models

In a two dimensional system with scalar order parameter (e.g., ferromagnetic Ising model), for $T < T_c$, the coexisting regions of $+$ and $-$ spontaneous magnetization are separated by a line interface. A macroscopic contour separating $+$ and $-$ spins, can in principle assume complicated geometric configurations. However, it is generally accepted that for studying wetting transitions and related interfacial fluctuation phenomena it is sufficient to account for the SOS subset of interface configurations (Fig. 3) *with no overhangs or bubbles.* Such configurations are specified as follows. Consider a planar square lattice of Ising spins, $\pm 1$, at positions $(X, Y)$ with integer $0 \leq X < \infty$, $|Y| < \infty$, i.e., a semi-infinite half plane geometry. The boundary spins are fixed at $-1$ for $X = 0$ and $+1$ for $X = \infty$, thus forcing an interface, which separates the region of predominantly $-$ magnetization near the wall at $X = 0$ from the region of $+$ magnetization for large $X$. In the SOS description of the interface (Fig. 3), we consider spin configurations with $n_Y \geq 1$ leftmost spins (at $X = 0, 1, \ldots, n_Y - 1$) in each fixed-$Y$ row taking values $-1$, while the remaining spins (at $X = n_Y, n_Y + 1, \ldots$) taking values $+1$. Here $Y = 0, \pm 1, \pm 2, \ldots$ labels the lattice rows: see Fig. 3.

Clearly, the SOS configurations of the interface are geometrically identical to the PDSAWs on the square lattice (with $\pm X$ and $+Y$ steps). However, in the SOS models, different interface configurations have different energies. The SOS interfacial energy is specified by the Hamiltonian

$$H/k_B T = \sum_Y [U \mid n_Y - n_{Y-1} \mid -W \delta_{1 n_Y} + E(n_Y)]. \qquad (3.1)$$

Here $U > 0$ models the surface tension contribution. Contact interactions attracting the interface to the wall at $X = 0$ are represented by the $W > 0$ term. The external-field and residual wall interactions are denoted by $E(n)$.

In (3.1), the difference $|n_Y - n_{Y-1}|$ can take on any value. However, it is mathematically convenient and physically acceptable to further restrict the model to configurations with $|n_Y - n_{Y-1}| = 0$ or 1, for all $Y$. Detailed studies indicate that such a *restricted* SOS model (RSOS) is identical with the unrestricted model in all the qualitative features of the wetting behavior.

It is convenient to introduce the notation

$$0 < u \equiv e^{-U} < 1, \quad w = e^{W - E(1)} > 1, \qquad (3.2)$$

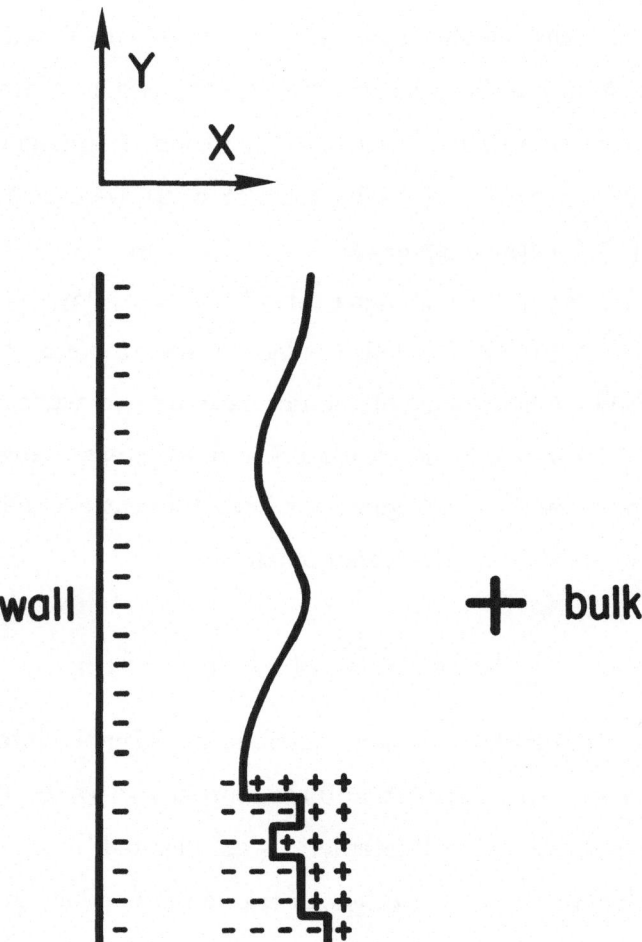

**Figure 3**

Solid-on-Solid modeling of Ising interface as a continuous structureless string (with no overhangs, etc.). The lower part illustrates the lattice representation as a directed walk of $\pm X$ and $+Y$ steps.

where we assume $W > E(1)$ to avoid unilluminating mathematical complications. [The physically interesting interactions are usually such that $W >> |E(1)|$.] Also, we denote by $n$ and $m$ the $n_Y$ values in two consecutive rows. Then the *transfer matrix* $T$ can be defined, with elements

$$T_{nm} = u^{|n-m|} w^{\delta_{1n}} e^{-E(n)} \left(\delta_{0,n-m} + \delta_{1,|n-m|}\right). \tag{3.3}$$

Note that we choose a nonsymmetric transfer matrix. Let $g_m$ denote the right–eigenvector elements. Then the eigenvalue equations

$$\sum_{m=1}^{\infty} T_{nm} g_m = \lambda g_n \tag{3.4}$$

reduce to

$$g_n + u\left(g_{n-1} + g_{n+1}\right) = \lambda g_n e^{E(n)}, \quad n > 1, \tag{3.5}$$

which is a second order difference equation, with the *boundary condition*

$$w\left(g_1 + u g_2\right) = \lambda g_1. \tag{3.6}$$

In order to illustrate the mechanism of the wetting transition, we will now solve (3.5)-(3.6) with no external potential, i.e., $E(n) \equiv 0$. It is convenient to introduce two new variables, $t$ and $\epsilon$, defined by

$$\frac{1}{u} = \frac{1}{u_c} - \frac{w}{w-1} t \quad \text{with} \quad u_c \equiv \frac{w-1}{2-w}, \tag{3.7}$$

$$\lambda = 1 + 2u + 2u\epsilon. \tag{3.8}$$

The general solution of (3.5) with $E(n) \equiv 0$ is

$$g_n = A\gamma^n + B\gamma^{-n} \qquad \text{for} \qquad \epsilon \neq 0, -2, \tag{3.9}$$

$$g_n = A\gamma^n + Bn\gamma^n \qquad \text{for} \qquad \epsilon = 0, -2, \tag{3.10}$$

where

$$\gamma \equiv 1 + \epsilon - \sqrt{\epsilon(2 + \epsilon)}. \tag{3.11}$$

For $\epsilon > 0$, we have $\gamma < 1$. On physical grounds we discard the exponentially growing term in (3.9), i.e., $B = 0$. The eigenvector $g_n \propto \gamma^n$ is then dominated by the "nonwet" spin configurations, with the layer of $-$ spins extending the distance

$$\xi_\perp = (-\ln \gamma)^{-1} \tag{3.12}$$

from the wall. However, the boundary condition (3.6) "quantizes" the nonwet part of the spectrum, yielding at most one eigenvalue. One can show that this nonwet solution exists only for $u < u_c(w)$, corresponding to $t < 0$ in (3.7).

The eigenvectors corresponding to $-2 \leq \epsilon \leq 0$ are dominated by the "wet" configurations with an unbounded $-$ layer. (The range $\epsilon < -2$ is of no physical interest.) The end points $\epsilon = 0, -2$ require special consideration, which we omit here except to quote that $B(\epsilon = 0) = 0$, $g_n(\epsilon = 0) \equiv A$. For $-2 < \epsilon < 0$, we note that $\gamma$ and $\gamma^{-1} \equiv \gamma^*$ become complex (and conjugate), with $|\gamma| = 1$. The boundary condition then determines the ratio $A/B$; the "wet" spectrum is not quantized. It exists for all $t$, covering the $\lambda$ range $1 - 2u \leq \lambda \leq 1 + 2u$.

The interfacial free energy $\sigma$ and the longitudinal correlation length $\xi_\parallel$ of the system are given, as usual in transfer matrix calculations, in terms of the largest and second eigenvalues $\lambda_0$ and $\lambda_1$, with the corresponding $\epsilon_{0,1}$ values, by

$$\sigma = -\ln \lambda_0, \qquad \xi_\parallel^{-1} = \ln (\lambda_0/\lambda_1) ; \tag{3.13}$$

one should also use $\gamma(\lambda_0)$ in (3.12) to obtain a definition of the transverse correlation length. For $t < 0$ $[u < u_c(w)]$, which corresponds to the nonwet regime, explicit calculation yields

$$\lambda_0 = \frac{w}{2} \left[ 1 - \sqrt{1 + \frac{4u^2}{w-1}} \right], \tag{3.14}$$

while $\lambda_1 \equiv 1 + 2u$. For small negative $t$, we find by expanding (3.14),

$$\epsilon_0 \approx \frac{1}{2} t^2. \tag{3.15}$$

Relations (3.7)-(3.8) can be used to obtain the following rather general small-$t$ and $\epsilon_0$ expansion for $\sigma$ in (3.13), valid to $O(t^2)$ and $O(\epsilon_0)$,

$$\sigma = \ln \frac{2-w}{w} - \frac{2(w-1)}{2-w} t - \frac{2(w-1)}{(2-w)^2} t^2 - \frac{2(w-1)}{w} \epsilon_0 + \dots . \tag{3.16}$$

Note that this expansion does not depend on the particular form of $\epsilon_0(t; w)$, e.g., (3.15). The *singular part* of the interfacial free energy is thus proportional to $-\epsilon_0$ and is given by

$$\sigma_{\text{sing}} = -\frac{w-1}{w} t^2 \tag{3.17}$$

for small negative $t$. (Note that the "regular part" of $\sigma$ has no obvious interpretation unless one takes special care to relate the SOS parameters to the original Ising model formulation.) For the correlation lengths, we use (3.12)-(3.13) to get

$$\xi_\parallel \approx \frac{w}{w-1} t^{-2}, \qquad \xi_\perp \approx |t|^{-1} . \tag{3.18}$$

As $t \to 0^-$, there is a wetting critical point corresponding to the depinning of the interface from the wall. For *positive* $t$, $\lambda_0 \equiv 1 + 2u$, with constant $g_n$, and the spectrum is continuous (gapless). Thus, formally we obtain

$$\sigma_{\text{sing}} = 0, \quad \xi_\| = \infty, \quad \xi_\perp = \infty, \tag{3.19}$$

in the wet regime. These results illustrate some basic features of wetting transition in SOS models with short-range forces, and in the absence of the external potential $E(n)$. (These properties are believed to be valid for all two dimensional models with scalar order parameter.)

In the more general case, the nature of the wetting transition in the SOS model depends on the form of the external potential $E(n)$. The following choices of $E(n)$ are of particular interest.

*Exponential short–range potentials,* behaving for large $n$ according to

$$E(n) \approx ce^{-An}, \quad A > 0 \quad (n \gg 1). \tag{3.20}$$

Such exponential potentials are generated in the process of renormalization of the wetting models with "contact" short-range forces.

*Power law long–range potentials,* behaving for large $n$ according to

$$E(n) \approx cn^{-\phi}, \quad \phi > 0 \quad (n \gg 1). \tag{3.21}$$

Such potentials are of practical importance in $3d$ wetting, and have been extensively studied for the $2d$ case.

*Applied field–like binding potentials,*

$$E(n) = cn^\psi, \quad c > 0, \quad \psi > 0. \tag{3.22}$$

Potentials of this form always suppress the wetting transition (bind the interface). However, one can study the $c \to 0^+$ scaling behavior. The choice $\psi = 1$, corresponding to the applied magnetic field, is of special interest and has been considered by many authors, within the differential equation approximation, see below.

In actual calculations, it is convenient to "de-exponentiate" the potential. For the long–range potentials of the type (3.21), one argues that the nature of the wetting transition depends mostly on the long–range tail while the short range features of the potential represent a perturbation of the contact, $W$, interaction in (3.1), provided $|E(n)| << W$ for $n = O(1)$. Thus one can choose a *power law potential*

$$E(n) \equiv \ln(1 + cn^{-\phi}), \qquad \phi > 0, \qquad c - \text{small}, \qquad (3.23)$$

which satisfies (3.21). In Section C we discuss an alternative choice of a *power law potential*,

$$E(n) \equiv \ln\left[1 + \frac{c}{n(n+1)\ldots.(n+\phi-1)}\right], \qquad (3.24)$$

for *integer* $\phi = 1, 2, \ldots$ (and small $c$).

A similar line of reasoning for the exponential potentials (3.20), is somewhat ambiguous since they are short–range all along. However, for mathematical convenience, the choice of the *exponential potential*

$$E(n) \equiv \ln(1 + ce^{-\mathcal{A}n}), \qquad \mathcal{A} > 0, \qquad (3.25)$$

is used, which allows derivation of analytic results, and a detailed analysis for small $c$ (see Section C).

"De-exponentiation" of the binding potentials (3.22), i.e., using *modified binding potentials*

$$E(n) = \ln(1 + cn^{\psi}), \qquad \psi > 0, \qquad c > 0, \tag{3.26}$$

obviously changes the large-$n$ asymptotic form. However, it can be shown that the *small-c* scaling behavior is not affected.

## B. Generating functions, differential equations, continued fractions

In the presence of non-contact interactions, the lattice SOS models are no longer exactly solvable, specifically (3.5) can not be solved for general $E(n)$. This section is devoted to various methods of obtaining approximate and in some cases exact results for the RSOS difference equation (3.5).

The *generating function method* for solving linear difference equations is well known. Here we emphasize features specific for the RSOS model applications. The generating function is defined by

$$G(z) = \sum_{n=1}^{\infty} g_n z^{n-1} = g_1 + g_2 z + g_3 z^2 + \dots. \tag{3.27}$$

Eq. (3.5) is then multiplied by $z^n$ and summed over $n = 2, 3, \dots$. In some cases, one ends up with a closed form equation for $G(z)$. Specifically, for potentials of the type (3.23), (3.24), (3.26), with *integer* $\phi, \psi$, one obtains differential equations for $G(z)$.

In order to illustrate the generating function approach, including the quantization of the "nonwet" eigenvalues imposed by the boundary condition (3.6), we turn again to the simple solvable case $E(n) \equiv 0$. [Here and below we will be mostly interested in the nonwet regime of finite $\sigma_{\text{sing}}$, $\xi_{\parallel}$ and $\xi_{\perp}$ , while (3.19) is typical for the wet phase.] The appropriate equation for $G(z)$ is then *algebraic*,

$$\left[u\left(1+z^2\right)+(1-\lambda)z\right]G(z)=\left[u+(1-\lambda)z\right]g_1+uzg_2. \tag{3.28}$$

By using (3.8) and (3.11), this can be represented as

$$G(z)=g_1\frac{1-z(2+2\epsilon-g_2/g_1)}{(z-\gamma)(z-\gamma^{-1})}. \tag{3.29}$$

The nonwet solution corresponds to $g_n\to 0$ for large $n$. Since $g_n$ are the Taylor coefficients of $G(z)$, we conclude that two conditions must be satisfied. First, $\epsilon>0$ is needed to have real $0<\gamma<1$. Secondly, the singularity at $z=\gamma$ yielding exponentially divergent $g_n$, must be cancelled. However, the ratio $g_2/g_1$ can be replaced by

$$\frac{g_2}{g_1}=\frac{w(1+t)+2\epsilon}{w}, \tag{3.30}$$

as implied by the boundary condition (3.6), with (3.7)-(3.8). Canceling the pole at $\gamma$ yields therefore the relation between $\epsilon$ and $t$ which determines the "quantized" eigenvalue $\epsilon_0(t;w)$, see (3.14)-(3.15). Note that since the original difference equations, (3.5) and (3.6), are linear, $G(z)$ has an arbitrary coefficient $g_1$, in (3.29).

In order to describe the "continuum" *differential equation* approach, let us introduce the notation

$$E(n)\equiv\ln\left[1+\frac{w-1}{w}V(n)\right], \tag{3.31}$$

which effectively defines $V(n)$ for $n=1,2,3,\ldots..$ After some algebra, equation (3.5) [with (3.7), (3.8), (3.31)] can be expressed as

$$- (g_{n+1} - 2g_n + g_{n-1}) + \left[ 1 + \frac{2(w-1)}{w} \epsilon - t \right] V(n) g_n = (-2\epsilon) g_n, \qquad n > 1. \tag{3.32}$$

A standard procedure for the critical region near the wetting transition (or rounded transition in the case of binding potentials), i.e., for *small t* and $\epsilon$, is to approximate (3.32) by the differential equation

$$- \frac{\partial^2 g(X)}{\partial X^2} + V(X) g(X) = (-2\epsilon) g(X), \tag{3.33}$$

where $0 \leq X < \infty$ is a continuous counterpart of $n$. Indeed, the fluctuations become large near the transition, and the magnetization profile varies over large distances (comparable to $\xi_\perp$). Thus, the discreteness of the original problem will be "washed out". [The small $O(t, \epsilon)$ terms have been discarded in the coefficient of the potential.]

The boundary condition (3.30) is written as

$$\frac{g_2 - g_1}{g_1} = t + \frac{2}{w} \epsilon, \tag{3.34}$$

and is replaced by

$$\frac{g'(0)}{g(0)} = t. \tag{3.35}$$

Neglecting the $\epsilon$ term is justified since for sharp continuous wetting transitions $\epsilon \sim |t|^{2-\alpha}$ with $\alpha < 1$. For first–order wetting transitions ($\alpha = 1$) and for binding potentials, more care may be required.

It should be emphasized that in defining the potentials and the $X$-coordinate displacements ($n$ and $X$), we assume that the lattice spacing is 1. In the continuous differential equation approximation, one may wish to introduce the lattice spacing $a$, so that the distance from the wall is actually $na$ (or $Xa$). The coefficients $c$ and $\mathcal{A}$ in the definitions (3.20)-(3.26) of the interaction potentials can then be appropriately related to the "physical" quantities which remain finite while $a \to 0$ in the continuous limit.

Relation (3.33) is a quantum mechanical Schrödinger equation with potential $V(X)$ and with the boundary condition which corresponds to a point–like attracting (for $t < 0$) delta–function potential at the origin, $2t\delta(X)$. The $V(X) = 0$ wetting transition corresponds to the disappearance of the bound state as $t \to 0^-$. [To make this interpretation precise, one should restrict consideration to even wave functions and extend the problem to $-\infty < X < \infty$ symmetrically, i.e., with $V(|X|)$]. We will return to (3.33)-(3.35) in Section C, in connection with potentials (3.24).

A more *ad hoc* approach is to define the $2d$ SOS model by the quantum mechanical (QM) Hamiltonian with a potential consisting of a hard wall at $X = 0$, followed by a potential well at, say, $0 < X < b$, and parameters adjusted to have one loosely bound state. Long–range potentials of the form (3.23), (3.26) are then introduced, i.e.,

$$V(X) = CX^{-\phi}, \qquad \phi > 0, \tag{3.36}$$

or

$$V(X) = CX^{\psi}, \quad \psi > 0, \quad C > 0, \tag{3.37}$$

with $C \equiv cw/(w - 1)$. Both the discrete and continuous SOS models in all their varieties supposedly approximate the original Ising problem to the extent of describing the wetting transition singularities. Thus, there is no *a priori* classification by the degree of approximation. We believe, however, that in the case of the first–order transitions, and for the description of interfacial pinning by the binding potentials deep in the wet regime ($t > 0$), the discrete models are more appropriate, and the physical interpretation of their parameters is more transparent.

Although we do not intend to review in detail all the QM results available in the literature, let us mention some conclusions of general interest. Consider first the power law potentials (3.36). For $\phi > 2$, the mechanism of the wetting transition by the disappearance of the bound state into the continuum is not changed. Nonanalytic corrections to scaling are present, e.g., in (3.15), however, the leading order dependence $\sigma_{sing} \sim -t^2$ remains unchanged (with a modified $t$). For $0 < \phi < 2$ and $c < 0$, the wetting transition is suppressed: the potential is strong enough to pin the interface to the wall. In the spectral language, there are always bound states in addition to the continuous spectrum. For $t > 0$, one finds $\sigma_{sing} \sim |c|^{2/(2-\phi)}$ for small $|c|$, up to possible logarithmic corrections for $t \simeq 0$. A detailed study of the $\phi = 1$ case has been reported for the discrete difference equation, see below. It transpires that the wetting transition is first–order for $c > 0$, with some unusual properties, e.g., divergent correlation lengths (at least for $\phi = 1$).

For the binding potentials (3.37), the spectrum is always discrete. There is no wetting transition. For $t > 0$, one has $\sigma_{sing} \sim c^{2/(2+\psi)}$. Detailed results including the $c, t \to 0$ crossover scaling forms for various quantities are available in some cases. Finally, in the borderline case $\phi = 2$ in (3.36), a rich phase diagram with nonuniversal critical, multicritical, and first–order transitions has been discovered.

A linear second order difference equation, like (3.5), can be solved *formally* in terms of *continued fractions* (this method will be used extensively in Chapter V). Thus, we introduce the ratios

$$R_n \equiv g_{n+1}/g_n \quad (n \geq 1), \tag{3.38}$$

so that $R_1$ is given by the right hand side of (3.30). The difference equation (3.32) is then divided by $g_n$ and after some rearrangement of terms expressed as

$$R_{n-1} = \cfrac{1}{2(1+\epsilon) + \left[1 + \frac{2(w-1)}{w}\epsilon - t\right]V(n) - R_n}, \quad n \geq 2. \tag{3.39}$$

This can be iterated to generate a continued fraction expansion for $R_k$ $(k \geq 1)$. Specifically, for $R_1$ we obtain [see (3.30)]

$$1 + t + \frac{2}{w}\epsilon = \tag{3.40}$$

$$\cfrac{1}{2(1+\epsilon) + \tilde{V}(2) - \cfrac{1}{2(1+\epsilon) + \tilde{V}(3) - \cfrac{1}{2(1+\epsilon) + \tilde{V}(4) - \cfrac{1}{2(1+\epsilon) + \tilde{V}(5) - \ldots}}}},$$

where

$$\tilde{V}(n) \equiv \left[1 + \frac{2(w-1)}{w}\epsilon - t\right]V(n).$$

Eq. (3.40) is a formal implicit equation for $\epsilon(t)$.

Generally, the second-order difference equation (3.32) has two linearly independent solutions, say, $g_n^{(1)}$ and $g_n^{(2)}$. There is some arbitrariness in selecting the two solutions. However, if one can select them in such a way that

$$\lim_{n \to +\infty} \left[ g_n^{(1)} / g_n^{(2)} \right] = 0, \tag{3.41}$$

then $g_n^{(1)}$ is termed the *minimal* solution. The existence of the minimal solution is not granted. Indeed, our calculations in the previous section for the case $E(n) \equiv 0$ corresponding to $V(n) \equiv 0$, indicate that the minimal solution exists for $\epsilon > 0$, in which case it is the physical "nonwet" solution $A\gamma^n$. However, none of the two solutions for $-2 < \epsilon < 0$ are minimal. [On the borderline, $\epsilon = 0$, the physical solution $g_n = const$ is minimal.] The differential equation approximation (see above) suggests that the *nonwet* solution being the minimal solution is a general rule. Indeed, it corresponds to the quantized localized ground state eigenfunction in QM calculations which decays at least exponentially as $X \to +\infty$. On the other hand, the eigenfunctions of the continuous spectrum are linear combinations of two running waves none of which is "minimal" as $X \to +\infty$.

An important theorem by Pincherle relates the convergence of the continued fraction for $R_1$ (and $R_k$ with $k > 1$) to the existence of the minimal solution. Indeed, the right hand side of (3.40), and similar continued fractions for $k > 1$, converge *if and only if* the difference equation possesses the minimal solution. Furthermore, the values of the continued fractions give $R_k$ for the minimal solution, i.e., $g_n = g_n^{(1)}$ in (3.41). Thus, (3.40) is a well defined equation for the ranges of the parameters $t, w$, and those of $V(n)$, for which the nonwet solution exists, and its free energy is given by the largest root $\epsilon_0(t; w, \ldots)$.

The line of the argument can be reversed. The general mathematical theory of the convergence of continued fractions can be invoked to make some of the conclusions on the spectrum of the problem more rigorous. For example, for the binding potentials (3.22), (3.26) corresponding to $V(n) \to \infty$ for large $n$, the appropriate type of continued fractions converge for all $\epsilon$. Thus the boundary conditions will quantize all "energies" $\epsilon$. However, for all other potentials introduced in Section A, with $V(n) \to 0$, one can prove that the continued fraction converges for $\epsilon > 0$ [and sometimes for $\epsilon = 0$, depending on the details of $V(n)$], but it diverges for $-2 < \epsilon < 0$. Thus, only the $\epsilon > 0$ part of the physically relevant spectrum is quantized and can represent nonwet solutions.

The continued fraction in (3.40) is of the type called a J–fraction in the mathematical literature. Unfortunately, not much is known about the analytic form of such fractions for $\epsilon \to 0^+$. In the case of potentials $V(n) \to 0$ (for large $n$), $\epsilon = 0$ is a special point since, as already mentioned, the J–fraction converges only for $\epsilon > 0$ or $\epsilon \geq 0$. This can be seen in the simplest case of no external potential. Indeed, for $V(n) \equiv 0$ the continued fraction is easily evaluated: for $\epsilon \geq 0$, it converges to $\gamma(\epsilon)$, see (3.11), and (3.40) reduces to

$$1 + t + \frac{2}{w}\epsilon = 1 + \epsilon - \sqrt{\epsilon(2 + \epsilon)}. \tag{3.42}$$

Thus the continued fraction, and $t(\epsilon)$, have a $\sim \sqrt{\epsilon}$ singularity as $\epsilon \to 0^+$. Note that (3.42) yields (3.14). Specifically, for small $\epsilon$, (3.42) is just $t \approx -\sqrt{2\epsilon}$ which has one nonwet solution (3.15), for $t < 0$ only. Generally, the continued fraction equation (3.40) becomes an algebraic equation for $\epsilon(t)$ if $V(n)$ is of finite range, i.e., $V(n > n_{\max}) = 0$.

## C. Exact solutions of RSOS models with external potentials

In this section we describe two analytic solutions of the RSOS model, for short- and long-range external potentials. First, consider the exponential potential

$$V(n) = Ce^{-An}, \qquad A > 0. \tag{3.43}$$

In the notation of Eqs. (3.25) and (3.31), $C = cw/(w-1)$. The difference equation (3.32) reads

$$g_{n+1} - 2(1+\epsilon)g_n + g_{n-1} = \left[1 + \frac{2(w-1)}{w}\epsilon - t\right] C e^{-An} g_n. \tag{3.44}$$

The difference equations of this type will be analyzed in detail in Chapter V. Here, we only indicate that the following series for the *minimal* solution (we keep $\epsilon > 0$ here) can be used,

$$g_n = \gamma^n \sum_{m=0}^{\infty} p_m e^{-Anm}, \tag{3.45}$$

where the coefficients $p_m$ satisfy the *first-order* recursion ($m > 0$),

$$\left[\gamma\left(e^{-Am} - 1\right) + \gamma^{-1}\left(e^{Am} - 1\right)\right] p_m = C\left[1 + \frac{2(w-1)}{w}\epsilon - t\right] p_{m-1}, \tag{3.46}$$

obtained by substituting (3.45) in (3.44). With a convenient choice $p_0 = 1$, we get

$$p_m = C^m \left[1 + \frac{2(w-1)}{w}\epsilon - t\right]^m \prod_{k=1}^{m} \left[\gamma^{-1}\left(e^{Ak} - 1\right) - \gamma\left(1 - e^{-Ak}\right)\right]^{-1}, \tag{3.47}$$

for $m = 1, 2, \ldots$. The boundary condition [see (3.30)] reduces to

$$1 + t + \frac{2}{w}\epsilon = \gamma \frac{\sum_{m=0}^{\infty} p_m e^{-2Am}}{\sum_{m=0}^{\infty} p_m e^{-Am}}. \tag{3.48}$$

This implicit equation for $\epsilon(t)$ is rather complicated for general $C$. However, the series on the right hand side are power series in $C$ since $p_m = O(C^m)$. Thus, for small $C$ a systematic expansion scheme can be developed by accounting for corrections due to successively higher powers of $C$. The leading singularity $(\sim \sqrt{\epsilon})$ of $t(\epsilon)$ still comes from the coefficient $\gamma$ on the right hand side of (3.48), see (3.11). Thus, the nature of the wetting transition is not changed (at least, for small $C$). For example, (3.15) is replaced by

$$\epsilon_0 \approx \left(\frac{1}{2} + \delta_c\right)(t - t_c)^2, \tag{3.49}$$

where the shifted $t$–variable can be calculated to a desired accuracy, as a power series in $C$. For example, to $O(C)$,

$$t_c \approx -C e^{-\mathcal{A}} \left(e^{\mathcal{A}} - 1\right)^{-1}, \tag{3.50}$$

$$\delta_c \approx C e^{-\mathcal{A}} \left(e^{\mathcal{A}} + 1\right) \left(e^{\mathcal{A}} - 1\right)^{-2}. \tag{3.51}$$

Let us now consider the class of power law, long–range potentials defined by (3.24) for mathematical convenience. Thus, we have

$$V(n) = \frac{C}{n(n+1)\dots(n+\phi-1)}, \tag{3.52}$$

where $\phi = 1, 2, 3, \dots$ and

$$C \equiv \frac{cw}{w - 1}. \tag{3.53}$$

Multiplication of the difference equation (3.32) by $Cz^{n-1}/V(n)$ and summation over $n = 2, 3, \dots$ yields after some algebra the following differential equation for the generating function [see (3.27)],

$$\frac{\partial^\phi}{\partial z^\phi} \left[z^{\phi-1} \left(z - \gamma\right)\left(z - \gamma^{-1}\right) G(z)\right] = \tag{3.54}$$
$$C\left[1 + \frac{2(w-1)\epsilon}{w} - t\right]\left[G(z) - g_1\right] + (\phi!)\left[g_2 - \left(\gamma + \gamma^{-1}\right) g_1\right],$$

where $\gamma(\epsilon)$ is defined by (3.11). The closed form solution of (3.54), with the additional condition (3.30), is known only for the case $\phi = 1$.

The inhomogeneous term in (3.54) is a constant. Thus, additional $z$–differentiation yields a homogeneous differential equation of order $\phi+1$. Small–$z$ analysis then indicates that out of its $\phi+1$ linearly independent solutions only *two* admit power series expansion around $z = 0$. The conditions $G(0) = g_1$ and $G'(0) = g_2$ then determine the coefficients of the linear combination of these two solutions. Thus, we end up with

$$G(z) = g_1 f_1(z; \epsilon, t, w, c) + g_2 f_2(z; \epsilon, t, w, c). \qquad (3.55)$$

For the wet regime, $-2 \le \epsilon \le 0$, the singularities of $f_k(z)$ nearest to the origin will be on the unit circle, in fact, at $z = \gamma$ and $\gamma^{-1} \equiv \gamma^*$ (where $|\gamma| = 1$). The ratio $g_2/g_1$ is fixed by (3.34), however, there is no quantization of $\epsilon$. In the nonwet regime, $\epsilon > 0$, the additional condition of cancelling, in $G(z)$, the singularity of $f_k(z)$ at $z = \gamma < 1$ to let the singularity at $z = \gamma^{-1}$ dominate the convergence of the power series (3.27), will lead to the quantization of $\epsilon$.

General expectations for $\phi = 1, 2, 3, \ldots$ presented above, have been checked in detail by exact calculations for the simplest case $\phi = 1$. The explicit form of $f_k(z)$ (not given here) involves hypergeometric functions. The resulting equation for $\epsilon_0(t; w, c)$ is rather complicated, and is not reproduced here. Scaling analysis for small $t, \epsilon$ and $c$ yields the following results. For $c < 0$ potentials, causing attraction of the interface to the substrate, the wetting transition is no longer sharp. The rounding is described asymptotically by the crossover scaling form

$$\epsilon_0 \approx \bar{c}^2 P\left(\frac{t - t_c}{\bar{c}}\right), \qquad (3.56)$$

where

$$\bar{c} = cw_0/(w_0 - 1), \text{ with } w_0 \equiv e^W, \tag{3.57}$$

and

$$t_c = \frac{w_0}{w_0 - 1}c\ln|c| + \left[1 + \frac{w_0}{w_0 - 1}\left(\mathcal{E} + \ln\frac{w_0}{w_0 - 1}\right)\right]c + o(c), \tag{3.58}$$

with Euler's constant $\mathcal{E} = 0.5772156649\ldots$. Details on the form of the scaling function $P$, and implications of (3.56), can be found in the references listed at the end of this chapter. Note, in particular, the logarithmic nonscaling shift in $t$ which is an unusual feature.

For $c > 0$ potentials, which repel the interface thus competing with the contact wall interaction, the wetting transition remains sharp. However, it becomes first–order, but with divergent correlation lengths. The scaling form (3.56) applies with a different scaling function $P$. However, $P \equiv 0$ for $t > t_c$, and $P$ vanishes linearly as $t \to t_c^-$: one finds

$$\epsilon_0 \approx \frac{3}{2}\bar{c}(t_c - t), \tag{3.59}$$

for small fixed $c > 0$. This is reminiscent of a first–order transition since the derivative $\partial\epsilon_0/\partial t$ is discontinuous at $t_c$. However, for the correlation lengths $\xi_{\parallel}$ and $\xi_{\perp}$ one finds a continuous divergence with new exponents $\nu_{\parallel} = 1$ and $\nu_{\perp} = \frac{1}{2}$ (different from $\nu_{\parallel} = 2$, $\nu_{\perp} = 1$ for $c = 0$). Here $\xi_{\parallel}$ is defined by the leading transfer matrix spectral gap, see (3.13), while $\xi_{\perp}$ is defined by the exponential tail of the decay of $g_n$ (which may lead to results different from the definition by moments, in this case).

## D. Finite-size effects for the wetting transition in two dimensions

So far we have considered wetting transition in the RSOS model on the semi-infinite two dimensional lattice. In this section we will analyze the effect of finite size on this transition. Specifically, we study the RSOS model as defined in Section A, but on the infinite strip of width $N$, i.e., $0 \le X \le N$. In addition, we add to the Hamiltonian (3.1) the contact potential $W$ which attracts the interface to the wall at $X = N$, similar to the one at the $X = 0$ wall. Throughout this section we will take $E(n) \equiv 0$. Thus, the transfer matrix $T$ has elements

$$T_{nm} = u^{|n-m|} w^{\delta_{1n} + \delta_{Nn}}. \tag{3.60}$$

The eigenvalue problem is solved in close analogy with the $N = \infty$ case. The difference equation for the eigenvector elements $g_n$ is again

$$g_n + u(g_{n-1} + g_{n+1}) = \lambda g_n, \quad \text{for} \quad 1 < n < N, \tag{3.61}$$

while the boundary conditions now read

$$\lambda g_1 = w(g_1 + u g_2); \quad \lambda g_N = w(g_N + u g_{N-1}). \tag{3.62}$$

The general solution of (3.61) is given by (3.9) or (3.10), with $A$ and $B$ determined by the boundary condition (3.62). This yields a linear homogeneous system from which, by equating the determinant to zero, we get the relation

$$(\lambda - w)\gamma^{-1} - wu = \pm\gamma^{N-3}[(\lambda - w)\gamma - wu]. \tag{3.63}$$

One solution to this equation is $\lambda = 1 + 2u$, corresponding to $\gamma = 1$. However, this is a spurious root leading, for general $N$, to the vanishing eigenvector $g_n \equiv 0$, *except* for $u \equiv u_c$ (bulk transition point), given by (3.7), with $g_n \equiv 1$. Using the notation introduced in (3.7)-(3.8), relation (3.63) can be expanded, to the leading order in $t$ and $\epsilon$, as

$$\sqrt{2\epsilon} + t = \pm e^{-2\sqrt{2\epsilon}N}(\sqrt{2\epsilon} - t). \tag{3.64}$$

Analysis of these equations (with $\sqrt{\epsilon} = +i\sqrt{-\epsilon}$ for negative $\epsilon$) involves long but straightforward algebra, and we only summarize the results here. Equations (3.64) admit solutions of the scaling form

$$\epsilon_k = N^{-2}\mathcal{G}_k(\zeta), \quad \zeta \equiv tN/\sqrt{2}. \tag{3.65}$$

There are at most two positive-$\mathcal{G}$ roots. However, there are unbounded number of negative-$\mathcal{G}$ roots with the corresponding $\epsilon_k < 0$ values condensing with density $\sim N^2$ to form the upper edge of the continuous spectrum which in the bulk limit covers the $\lambda$ range $[1 - 2u, 1 + 2u]$.

The functions $\mathcal{G}_0(\zeta)$ and $\mathcal{G}_1(\zeta)$ corresponding to the two largest-$\lambda$ roots are given implicitly by the relations

$$\zeta = -\sqrt{\mathcal{G}_0}\tanh\sqrt{\mathcal{G}_0/2} \quad \text{for} \quad \zeta \leq 0,\ \mathcal{G}_0 \geq 0; \tag{3.66}$$

$$\zeta = \sqrt{-\mathcal{G}_0}\tan\sqrt{-\mathcal{G}_0/2} \quad \text{for} \quad \zeta \geq 0,\ -\frac{\pi^2}{2} < \mathcal{G}_0 \leq 0; \tag{3.67}$$

$$\zeta = -\sqrt{\mathcal{G}_1}\coth\sqrt{\mathcal{G}_1/2} \quad \text{for} \quad \zeta \leq -\sqrt{2},\ \mathcal{G}_1 \geq 0; \tag{3.68}$$

$$\zeta = -\sqrt{-\mathcal{G}_1}\cot\sqrt{-\mathcal{G}_1/2} \quad \text{for} \quad \zeta \geq -\sqrt{2},\ -2\pi^2 < \mathcal{G}_1 \leq 0. \tag{3.69}$$

The free energy scaling function $\mathcal{G}_0(\zeta)$ is analytic at the origin [as is $\mathcal{G}_1(\zeta)$] and can be expanded in the power series $\mathcal{G}_0(\zeta) = -\sqrt{2}\zeta + \frac{1}{6}\zeta^2 + \dots$ . For $\zeta \to \infty$, corresponding to $t \gg N^{-1}$, i.e., to the "wet" edge of the critical region, we have $\mathcal{G}_0(\zeta) \approx -\pi^2/2$. Thus, the singular part of the free energy vanishes according to $|\epsilon| \sim N^{-2}$, which is typical of the critical "soft mode" phases. For $\zeta \to -\infty$, $-t \gg N^{-1}$, in the "nonwet" regime we find $\mathcal{G}_0(\zeta) \approx \zeta^2$ up to exponentially small corrections. Note that in the limit $|\zeta| \to \infty$ the bulk critical behavior $\epsilon_0^{(bulk)} \approx t^2/2$ or 0, for $t \leq 0$ or $t \geq 0$ is recovered, see (3.15). The correlation length scaling follows from (3.13),

$$\xi_{\parallel} \approx \frac{w}{2(w-1)} N^2 [\mathcal{G}_0(\zeta) - \mathcal{G}_1(\zeta)]^{-1}. \tag{3.70}$$

The difference $\mathcal{G}_0 - \mathcal{G}_1$ is finite for all $\zeta \geq 0$, and in the limit $\zeta \to +\infty$, suggesting that $\xi_{\parallel} \sim N^2$. At $\zeta = 0$ this is the expected scaling result since the exponent 2 must be $\nu_{\parallel}/\nu_{\perp}$ for which the bulk values are $\nu_{\parallel} = 2$ and $\nu_{\perp} = 1$. The $N^2$ divergence for large positive $\zeta$ is another characteristic feature of the critical "soft mode" wet phase. In the "nonwet regime" limit $\zeta \to -\infty$, however, we find $\mathcal{G}_0 - \mathcal{G}_1 \approx 8\zeta^2 e^{\sqrt{2}\zeta}$. Thus, for $-t \gg N^{-1}$, $\xi_{\parallel} \sim t^{-2} \exp(|t|N)$. The exponential divergence of the correlation length is in accord with the general expectations for finite-size strips with nearly broken symmetry.

The *bulk* perpendicular correlation length $\xi_{\perp}^{(bulk)}$ is given by the spatial variation of the eigenvector entries $g_n$. Indeed, the identification $\gamma^{\pm n} \sim \exp(\mp n/\xi_{\perp})$ leads to $\xi_{\perp}^{(bulk)} \equiv (-\ln\gamma)^{-1}$, see (3.12). The critical behavior of this quantity is given in (3.18), i.e., $\xi_{\perp}^{(bulk)} \approx |t|^{-1}$, for $t < 0$. Thus, the finite-size scaling variable $|\zeta| \propto |t|N$ can be identified with $N/\xi_{\perp}^{(bulk)}$, at least in the nonwet scaling regime. Finally, we should note that exponentially diverging correlation length $\xi_{\parallel}$ exists for all $0 < u < u_c^-$, outside the critical region. Examination of (3.63) gives $\xi_{\parallel} \sim \exp[N/\xi_{\perp}^{(bulk)}]$, valid as long as $u_c - u \gg N^{-1}$. The existence of this length is reminiscent of the similar phenomenon observed with finite-size scaling in directed random walks (Chapter II). In the present case, the exponentially diverging length scale corresponds to the $Y$-distance between the "tunnelings" of the bound interface from one wall to another.

## E. Finite-size effects for fluctuating interfaces

In this section we explore finite-size scaling properties of unbound fluctuating interfaces. In order to make the problem somewhat more general, we consider *inclined* (tilted) interfaces. The model is specified as follows.

Consider a planar square lattice of Ising ($\pm 1$) spins located at lattice points $(X, Y)$, with $X = 0, 1, 2, ..., L$, and $Y = \pm\frac{1}{2}, \pm\frac{3}{2}, ..., \pm(M - \frac{1}{2})$, as shown in Fig. 4. By virtue of the boundary conditions, the SOS contour begins at $(0, 0)$ and ends at $(L, m)$, see Fig. 4, so that the average inclination of the interface can be measured by the angle $\theta$, given by $\tan\theta = m/L$. The SOS interface configurations are uniquely specified by the set of interface heights $h_i$, measured from the reference level $Y = 0$, as shown in Fig. 4. The heights can take integer values

$$h_i = 0, \pm 1, \pm 2, ..., \pm(M - 1), \quad \text{for} \quad i = 1, 2, ..., L - 1, \qquad (3.71)$$

with the additional requirement that $h_0 = 0$ and $h_L = m$. The SOS interfacial energy is specified by the Hamiltonian

$$H/k_B T = U \sum_{i=1}^{L-1} |h_{i+1} - h_i| + (L+1)V, \qquad (3.72)$$

where, as in (3.1), $U > 0$ is the surface tension contribution, and we have included the constant term $V(L + 1)$ which accounts for the number of times the interface cuts across the vertical bonds (each time a contribution $V$ is added). Since there are no pinning potentials, this interface is free to fluctuate. By the

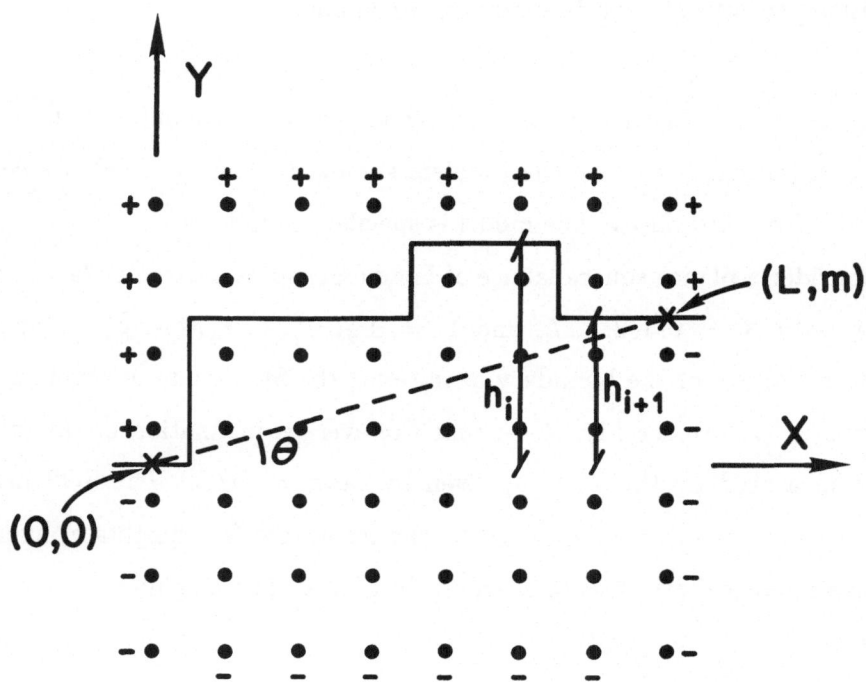

**Figure 4**

Two dimensional Ising model with inclined interface pinned by its end-points $(0,0)$ and $(L,m)$. The average inclination angle is specified by $\tan\theta = m/L$. A typical SOS interface configuration is shown by the solid line and is uniquely specified by the height variables $h_i$.

use of the transfer matrix technique, the partition function, $Z_{SOS}(m, L; M) \equiv \sum_{\{h_i\}} \exp(-H/k_B T)$, can be evaluated exactly for both restricted and unrestricted SOS models (interested reader should consult references listed in Section F). However, the full answer, which involves hypergeometric functions, is not illuminating. Thus, we quote below the final results appropriate for the finite-size analysis.

It turns out that for $m \ll L$, i.e., for small inclination angles $\theta$, and neglecting finite-$M$ effects (which are exponentially small for $M = O(L)$), the partition function can be written in the characteristic Gaussian form,

$$Z_{SOS}(m, L; M = \infty) \simeq \exp(-\tau_L L) \left(\frac{\kappa_L}{2\pi L}\right)^{1/2} \exp\left(-\frac{\kappa_L L \theta^2}{2}\right). \qquad (3.73)$$

Note that (3.73) is just the probability distribution for a directed random walk to start from the origin and reach the point $(L, m)$. The distribution is Gaussian in $\theta$, as seen in (3.73), with width $\sim 1/\sqrt{L}$. (This random walk property of line interfaces in two dimensions has found extensive uses in the literature.)

The leading finite-size contributions for $\tau_L$ and $\kappa_L$ are obtained explicitly as the standard $O(1/L)$ "endpoint" corrections,

$$\tau_L = \tau + \frac{a}{L} + o\left(\frac{1}{L}\right), \quad \kappa_L = \kappa + \frac{b}{L} + o\left(\frac{1}{L}\right), \qquad (3.74)$$

where $\tau$ and $\kappa$ are *bulk* surface tension and surface stiffness coefficient, respectively, while $a$ and $b$ are some finite-size coefficients. With $u = \exp(-U)$ and $v = \exp(-V)$, we have the explicit results

$$\tau = -\ln[v(1 + 2u)], \quad \kappa = \frac{1 + 2u}{2u} \quad \text{(RSOS model)}, \qquad (3.75)$$

$$\tau = -\ln\left[v\left(\frac{1 + u}{1 - u}\right)\right], \quad \kappa = \frac{(1 - u)^2}{2u} \quad \text{(SOS model)}, \qquad (3.76)$$

and

$$a = -\ln v, \quad b = \frac{1 - 2u - 8u^2}{8u^2} \quad \text{(RSOS model)}, \tag{3.77}$$

$$a = -\ln v, \quad b = \frac{(1 - u)^2(1 + 4u + u^2)}{8u^2} \quad \text{(SOS model)}. \tag{3.78}$$

If the inclination angle $\theta$ is fixed by the requirement $m = L \tan \theta$, then the *anisotropic* surface tension (more precisely, interfacial free energy), per unit length, in units of $k_B T$, is obtained from

$$\sigma(\theta, L; M) = -\frac{\cos \theta}{L} \ln Z_{SOS}(L \tan \theta, L; M), \tag{3.79}$$

which in $2d$ is an even function of $\theta$. Using (3.73)-(3.78), the following finite-size expressions for the surface tension and its second derivative are obtained

$$\sigma(0, L; M) = \sigma(0, \infty; \infty) + \frac{\ln L}{2L} + \frac{a - \ln \sqrt{\kappa/2\pi}}{L} + o\left(\frac{1}{L}\right), \tag{3.80}$$

and

$$\sigma''(0, L; M) = \sigma''(0, \infty; \infty) - \frac{\ln L}{2L} + \frac{b - a + \ln \sqrt{\kappa/2\pi}}{L} + o\left(\frac{1}{L}\right). \tag{3.81}$$

Note that the bulk quantities are generally defined via

$$\frac{\sigma(\theta, \infty; \infty)}{\cos \theta} = \tau + \frac{1}{2}\kappa\theta^2 + O(\theta^4), \tag{3.82}$$

with

$$\tau \equiv \sigma(0, \infty; \infty) > 0, \quad \kappa \equiv \sigma(0, \infty; \infty) + \sigma''(0, \infty; \infty) > 0. \tag{3.83}$$

Results (3.73), (3.80)-(3.81) hold generally within the capillary-wave theory. They are explicitly verified by the SOS model calculations. Note in particular the leading $\pm(\ln L)/2L$ corrections in (3.80)-(3.81) which contain no free parameters. Finally, we mention that the model just analyzed can be solved exactly for the finite-size corrections in its full lattice Ising version. The results are similar to (3.80)-(3.81).

## F. Selected literature

Recently, several comprehensive reviews of the wetting transitions and related surface phenomena have been published. The following articles are general surveys of the field.

**S. Dietrich,** in *Phase Transitions and Critical Phenomena*, Vol. 12,
    edited by C. Domb and J.L. Lebowitz (Academic, New York, 1988)

**M.E. Fisher,** J. Chem. Soc., Faraday Trans. *II*, **82**, 1569 (1986)

**D. Sullivan** and **M.M. Telo da Gama,** in *Fluid and Interfacial Phenomena,*
    edited by C.A. Croxton (Wiley, New York, 1986)

**P.-G. de Gennes,** Rev. Mod. Phys. **57**, 827 (1985)

The following reviews put emphasis on wetting transitions in the *two dimensional* Ising and SOS models.

V. **Privman** and **N.M. Švrakić,** J. Statist. Phys. **51**, 1111 (1988)

**D.B. Abraham,** in *Phase Transitions and Critical Phenomena*, Vol. 10,
     edited by C. Domb and J.L. Lebowitz (Academic, New York, 1986)

Results on fluctuations of unbound $2d$ interfaces, as well as on wetting, in finite-size geometries, have been collected from the following recent works: M.P. Gelfand and M.E. Fisher, Int. J. Thermophysics, *in print* (1989); V. Privman, Phys. Rev. Lett. **61**, 183 (1988); V. Privman and N.M. Švrakić, Phys. Rev. B**37**, 3713 and 5974 (1988); N.M. Švrakić, V. Privman and D.B. Abraham, J. Statist. Phys. **53**, 1041 (1988).

# IV. POLYMERS AT SURFACES

The behavior of a single polymer chain near an attractive surface is a subject of considerable experimental and theoretical importance. Such a chain may undergo adsorption-desorption transition from the state when it is mostly attached to the surface, to the state of detachment when the temperature is increased. This behavior finds applications in lubrication, adhesion, surface protection, etc. By methods similar to the techniques described in Chapters II and III, one can solve directed random walk models of polymer chains near surfaces (rigid or penetrable) and study the corresponding transition. In Section A, the relevant models are defined, and the adsorption-desorption transition at an impenetrable surface is analyzed exactly for two and three dimensional systems. Section B is devoted to a similar problem near a symmetric penetrable surface. Finally, Section C lists selected literature.

## A. Directed models of polymer adsorption

In order to study the statistical properties of polymer chains, we model their configurations by self-avoiding random walks (SAWs) on a lattice. (We consider here both the 2$d$ square and 3$d$ simple cubic lattices, and take, for simplicity, the lattice spacing to be unity.) Thus, the polymer chain consists of random self-avoiding steps (monomers) of unit length. To make this model exactly solvable, we impose *directedness* along the $+X$ axis, i.e., steps in the $-X$ direction are not allowed (this makes the walk partially directed). The attracting surface or wall, is positioned at the $X$ axis, for the 2$d$ model, and at the $XZ$ plane for the 3$d$ model. In the case of an impenetrable surface (considered in this section) the polymer can not extend below the $X$ axis (or $XZ$ plane in $3d$), i.e., only steps at $Y \geq 0$ are allowed. We consider a polymer chain which is attached by one end to the origin (grafted polymer), while the other end can be either pinned to the surface or free (dangling).

Consider a walk with the total of $N$ steps, $N_s$ of which are at the surface (i.e., $+X$, or $\pm Z$ steps at $Y = 0$). All allowed steps are assigned statistical weight or *fugacity* $z$. Furthermore, the steps at the surface have the total *excess* surface energy $E_s = -\Omega_s N_s$, with typically $\Omega_s \geq 0$, i.e., to each surface step we assign additional weight

$$w = \exp(-E_s/k_B T N_s) = e^{\Omega_s/k_B T} \geq 1. \tag{4.1}$$

The thermodynamic behavior of the chain is obtained from the partition function

$$Z(z, w) = \sum_{\text{all walks}} z^N w^{N_s}. \tag{4.2}$$

As usual, the average length of the walk is obtained from

$$\langle N(z, w) \rangle = \frac{\sum N z^N w^{N_s}}{\sum z^N w^{N_s}} \equiv z \frac{\partial \ln Z(z, w)}{\partial z}. \tag{4.3}$$

Similarly, the average number of monomers *at the surface* is

$$\langle N_s(z, w) \rangle = \frac{\sum N_s z^N w^{N_s}}{\sum z^N w^{N_s}} \equiv w \frac{\partial \ln Z(z, w)}{\partial w}. \tag{4.4}$$

The quantity of physical interest is the *fraction of adsorbed monomers*, $C_s(w)$, i.e., the fraction of the chain segments at the surface in the infinite-chain limit. This is given by

$$C_s(w) = \lim_{z \to z_\infty(w)} \frac{\langle N_s(z, w) \rangle}{\langle N(z, w) \rangle}, \tag{4.5}$$

where $z_\infty(w)$ is the fugacity value for which $\langle N \rangle \to \infty$. For simplicity, we describe the calculations for the $2d$ case, and only quote the $3d$ results, which can be obtained by a straightforward generalization.

In order to calculate the partition function (4.2) we use the transfer matrix approach as described in Chapter III; indeed, our model in $2d$ is quite similar to the SOS model of line interfaces. The configurations of the chain are uniquely specified by the heights $h_i$ (i.e., $Y$-values) above the surface, with $i = 0, 1, 2, ....$ The problem can be further simplified if the restriction

$$|h_i - h_{i+1}| = 0 \quad \text{or} \quad 1, \tag{4.6}$$

is imposed, i.e, at most a single $\pm Y$ step is allowed between any two $+X$ steps. This corresponds to the *restricted* model, similar to the RSOS model. Consider a partition function, $Z_L$, for walks with exactly $L$ steps in the positive $X$ direction, i.e., the walk ending at $X = L$. The total partition function (4.2) is then obtained as

$$Z = Z_0 + \sum_{L=1}^{\infty} Z_L, \tag{4.7}$$

where the term $Z_0$ accounts for the walks with no steps in the $+X$ direction, $Z_0^{(U)} = z/(1-z)$ or $Z_0^{(R)} = z$ for the unrestricted (U) or restricted (R) models. When $L \geq 1$, in the case of the unrestricted model, we have

$$Z_L^{(U)} = z^L \sum_{\{h_i\}=0}^{\infty} \delta_{0,h_0} z^\Xi w^\Upsilon, \tag{4.8}$$

with

$$\Xi \equiv \sum_{i=0}^{L} |h_{i+1} - h_i|, \quad \text{and} \quad \Upsilon \equiv \sum_{i=1}^{L} \delta_{0,h_i}. \tag{4.9}$$

The delta-term in (4.8) ensures that one end of the chain is pinned at the origin. The pinning of both ends can be accomplished by the additional factor $\delta_{0,h_{L+1}}$ in (4.8). Here $\Xi$ represents the total length of all vertical steps, whereas $\Upsilon = N_s$ for a given configuration. The partition function $Z_L$ in (4.8) can be evaluated by the transfer matrix method. Specifically, we define the transfer matrix $T$, with elements

$$T_{nm} = z^{|n-m|} w^{\delta_{0,n}}. \tag{4.10}$$

Then the summand in (4.8) can be represented as

$$\delta_{0,h_0} T_{h_1 h_0} T_{h_2 h_1} ... T_{h_{L+1} h_L} w^{-\delta_{0,h_{L+1}}}. \tag{4.11}$$

Therefore, we have

$$Z_L^{(U)} = z^L (\mathcal{R}^{(t)} T^{L+1} \mathcal{L}), \tag{4.12}$$

where the column vectors $\mathcal{R}$ and $\mathcal{L}$ account for the end effects, while the superscript $(t)$ in (4.12) denotes the transpose. The vector $\mathcal{L}$ has entries $\mathcal{L}_0 = 1$, $\mathcal{L}_{m>0} = 0$ (for the pinned end of the chain). For $\mathcal{R}$, we have $\mathcal{R}_0 = w^{-1}$, $\mathcal{R}_{m>0} = 1$ (for the dangling end). If both ends of the chain are pinned, one must use $\mathcal{R}_{m>0} = 0$. In the case of the restricted model, the formulation is the same, except that one uses

$$T_{mn}^{(R)} = (\delta_{0,n-m} + \delta_{1,|n-m|}) z^{|n-m|} w^{\delta_{0,n}}, \tag{4.13}$$

instead of (4.10). Clearly the transfer matrices with elements (4.10) and (4.13) are identical with the corresponding matrices for the SOS and RSOS models analyzed in Chapter III. However, the thermodynamic ensemble and the partition function

$Z$ are different. Namely, substituting (4.12) in (4.7) and omitting the unimportant term $Z_0$, we get

$$Z = z[\mathcal{R}^{(t)}T^2(1 - zT)^{-1}\mathcal{L}],\tag{4.14}$$

for both unrestricted and restricted models. Let us denote by $\lambda$ the eigenvalues of $T$. Then the partition function (4.14) becomes singular for $z\lambda_{max} = 1$, where $\lambda_{max}$ is the largest eigenvalue of $T$. This equation defines $z_\infty(w) < 1$. As discussed in Chapter III, the spectrum for both models consists of a continuous band of eigenvalues, with "plane wave" eigenvectors, and at most one bound state (bs) eigenvalue which corresponds to the eigenvector decaying as $\exp(-\varpi n)$. The values of $\lambda_{bs}$ and $\varpi$ are obtained from the conditions

$$\cosh \varpi = \frac{\lambda(1 + z^2) - (1 - z^2)}{2z\lambda}, \quad \lambda(1 - ze^{-\varpi}) = w(1 - z^2), \quad \text{(U)},\tag{4.15}$$

and

$$\cosh \varpi = \frac{\lambda - 1}{2z}, \quad \lambda - w = wze^{-\varpi}, \quad \text{(R)},\tag{4.16}$$

for the unrestricted and restricted models, respectively. The solution for $\lambda_{bs}$, with $\varpi > 0$, exists provided $w > w_*(z)$, where

$$w_*^{(U)}(z) = \frac{1}{1 - z}, \quad w_*^{(R)}(z) = \frac{1 + 2z}{1 + z}.\tag{4.17}$$

The continuous spectrum exists for all $w$, giving $\lambda$ in the range $(1 - z)/(1 + z) \leq \lambda^{(U)} \leq (1 + z)/(1 - z)$, and $1 - 2z \leq \lambda^{(R)} \leq 1 + 2z$.

As long as $w > w_*$, the bound state eigenvalue $\lambda_{bs}$, calculated from (4.15) or (4.16), is above the upper edge $\lambda_{cont}$ of the continuous spectrum, where $\lambda_{cont}^{(U)} = (1 + z)/(1 - z)$, $\lambda_{cont}^{(R)} = 1 + 2z$. In this regime, the equation $z\lambda_{max} = 1$ for $z_\infty(w)$ reduces to $z\lambda_{bs} = 1$, i.e.,

$$w(1 - z^2)(1 + z - zw) = 1, \quad \text{(U)}, \tag{4.18}$$

and

$$(w - 1)(1 - wz) = w^2 z^4, \quad \text{(R)}, \tag{4.19}$$

for the unrestricted and restricted models, respectively. The intersection of the curve $w = w_*(z)$, see (4.17), with (4.18) or (4.19) defines the *critical point* $(z_c, w_c) = (\sqrt{2} - 1, 1 + 1/\sqrt{2})$ or $(1/2, 4/3)$, for the unrestricted and restricted models, respectively.

For $w > w_c$, one finds that the equation $z\lambda_{max} = 1$ is satisfied in the regime where it reduces to $z\lambda_{bs} = 1$, and a nontrivial function $z_\infty(w)$ is given implicitly by (4.18) or (4.19). However, for $w \le w_c$, it turns out that $z\lambda_{max} = 1$ is solved in the regime where it is represented by $z\lambda_{cont} = 1$, and $z_\infty(w) \equiv z_c$ is constant. This phase diagram is shown in Fig. 5.

The values $z_\infty(w)$ correspond to chains with $\langle N \rangle = \infty$, as reflected by a singularity in $Z(z, w)$, for fixed $w$. Specifically, as $z \to z_\infty^-(w)$ for $w > w_c$, we have

$$Z(z, w) \sim \frac{1}{z_\infty(w) - z}. \tag{4.20}$$

As $w \to w_c$ from above, we have $z_\infty(w) \to z_c$ with *zero slope*: see Fig. 5. By using (4.20), with (4.3)-(4.5), we conclude that both $\langle N(z, w) \rangle$ and $\langle N_s(z, w) \rangle$ diverge $\sim [z_\infty(w) - z]^{-1}$, while the fraction of adsorbed monomers is given by

$$\mathcal{C}_s(w) = -\frac{w}{z_\infty(w)} \frac{dz_\infty(w)}{dw}. \tag{4.21}$$

As $w \to w_c^+$, the fraction of adsorbed monomers vanishes linearly. The condition $w > w_c$ specifies the adsorbed regime. For large values of $w$ (strongly attractive

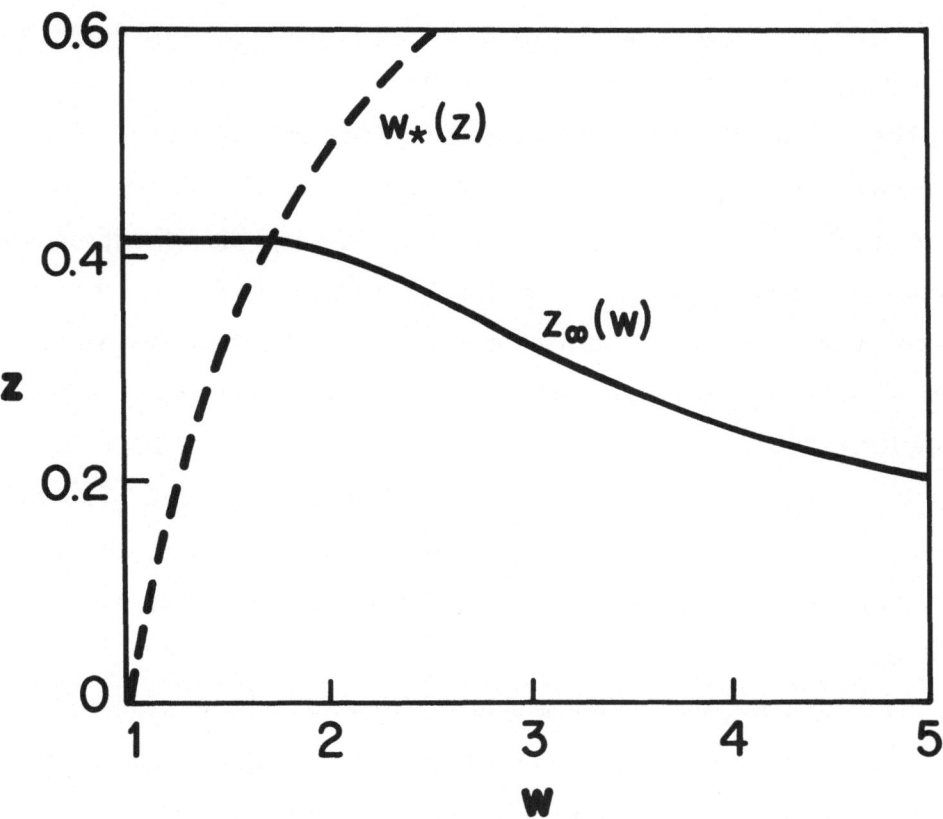

**Figure 5**

Adsorption-desorption phase diagram for the unrestricted model in 2*d*.

surface, or low temperature), $C_s(w)$ can be expanded in powers of $1/w$,

$$C_s(w) = 1 - \frac{3}{w^3} - \frac{4}{w^4} + ..., \tag{4.22}$$

for both models. The results for the two models differ in $O(w^{-5})$. The fraction of adsorbed monomers is shown in Fig. 6. For $w \leq w_c$, we have $z_\infty(w) \equiv z_c$, yielding $C_s(w \leq w_c) \equiv 0$. This corresponds to the desorbed polymer phase.

The behavior of the partition function near the singularity at $z_\infty(w)$ defines the exponent $\gamma$ [compare (2.7)]. Specifically, for $w > w_c$, $Z(z, w)$ behaves according to (4.20) as $z \to z_\infty^-$, giving $\gamma_1 = 1$ and $\gamma_{11} = 1$, where the subscript indicates whether the chain is pinned to the surface by one or both ends. The behavior is different when $z \to z_\infty^- = z_c^-$ at fixed $w < w_c$. The appropriate analysis involves spectral decomposition of the transfer matrix and is not detailed here. We only quote the results: $\gamma_1 = 1/2$ and $\gamma_{11} = -1/2$. Finally, at the borderline (i.e., $w = w_c$), one has the multicritical values $\gamma_1 = 1$ and $\gamma_{11} = 1/2$. The surface scaling relation, $2\gamma_1 - \gamma_{11} = \gamma + \nu_\perp$, is satisfied for both $w < w_c$ and $w = w_c$. [Recall that for directed walks without the surface, the bulk exponents $\gamma$ and $\nu_\perp$ are 1 and 1/2, respectively.]

For $3d$ systems, the analysis is similar. However, only the *restricted* model has been solved exactly. Since the unrestricted and restricted models in $2d$ have almost identical behavior, we anticipate that the description within the restricted $3d$ model should suffice. [The restriction is that there is at most one $\pm Y$ or $\pm Z$ step between any two $+X$ steps.] We omit the details of this calculation and only quote the results for the fraction of adsorbed monomers. One finds that $C_s(w)$ vanishes linearly as $w \to w_c = (23 - \sqrt{17})/16$ from above. When $w$ is large, the $(1/w)$-expansion yields

$$C_s(w) = 1 - \frac{2}{w} + \frac{12}{w^2} - \frac{83}{w^3} + ..., \tag{4.23}$$

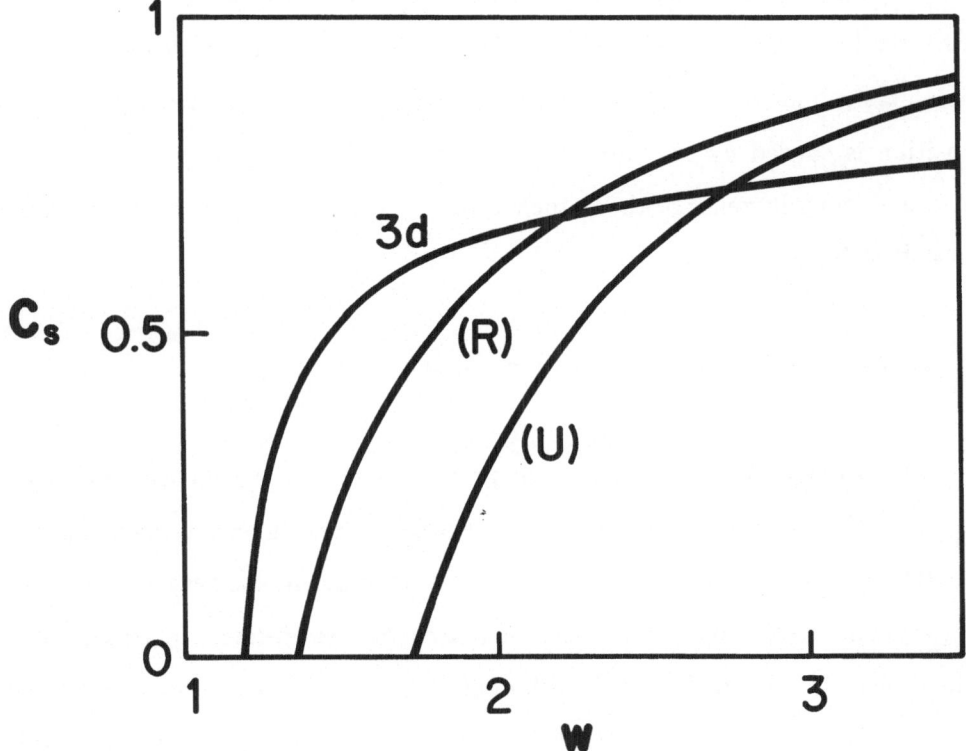

**Figure 6**

The fraction of adsorbed monomers for the unrestricted (U) and restricted (R) models in $2d$, as well as for the $3d$ model.

which should be compared with (4.22). Clearly, the adsorption for large $w$ is much weaker than in the $2d$ case, which is probably due to the large "phase space" available to the fluctuating chain. The behavior of the fraction of adsorbed monomers for the full range of parameters is shown in Fig. 6.

In summary, the behavior of polymers near surfaces in $2d$ and $3d$ systems is qualitatively similar, for directed models, exhibiting adsorption-desorption transition at the model-dependent values of surface interaction parameters. This transition is caused by the competition between the energy gain in adsorption, and the entropy increase of the detached polymer, and can be driven by temperature: Recall that $w = \exp(\Omega_s / k_B T)$.

## B. Polymer chains at penetrable surfaces

Using the methods described in the previous section, one can also analyze the behavior of the polymer chain near a *penetrable* surface. Physically, such a surface corresponds, e.g., to an interface between two media. In this section we assume the polymer interactions with these media to be *symmetric*. In experimental realizations, the two media can be different (asymmetric), and then the behavior near the adsorption-desorption transition is generically the same as that studied in the previous section. However, the symmetric case is interesting because there is no loss in entropy when a grafted chain is adsorbed at the surface. One thus anticipates that the depinning transition in this case can not be induced by the temperature increase, but only by making the surface repulsive, i.e., we anticipate that $w_c \equiv 1$.

Consider a directed random walk model of polymer chains on a $2d$ or $3d$ lattice, as described in the previous section. The surface is defined at the $X$ axis in $2d$ ($XZ$ plane in $3d$). The walk starts at the origin (grafted polymer) and is directed along $+X$ axis. We allow steps on *both* sides of the surface. Let $N$ be the total number of steps in the walk, $N_s$ of which are at the surface. In the grand canonical ensemble, the statistical weight (fugacity), $z$, is assigned to each step of the walk. Furthermore, steps *at* the surface are assigned additional weight $w$. Note that $w > 1$ correspond to attraction, while $w < 1$ would describe a repulsive surface. For such a walk the partition function is given by (4.2). Similarly, the definitions of the average length of the walk, the average number of surface steps, and the fraction of adsorbed monomers, are given by (4.3)-(4.5) respectively.

The solution of this model can be obtained by the transfer matrix method in close analogy with the impenetrable surface case. Throughout this section we will only consider the restricted model. For the $2d$ case, the transfer matrix is of the form (4.13), and the partition function is calculated from (4.14). However, the indices $m, n$ in (4.13) are no longer restricted to nonnegative values. Mathematically, the emergence of the singularity at $z_\infty(w)$ is the same as for the impenetrable surface. However, we will see below that $w_*(z) \equiv w_c = 1$.

The spectrum of the transfer matrix is similar to the one described in the previous section, with the bound state eigenvector now decaying as $\exp(-\varpi|n|)$, with $\varpi > 0$. The conditions for the existence of the bound state are

$$\cosh \varpi = \frac{\lambda - 1}{2z}, \tag{4.24}$$

and

$$\lambda - w = 2wze^{-\varpi}, \tag{4.25}$$

which differs from (4.16) by the factor of 2. The bound state eigenvalue exists *provided* $w > w_* \equiv 1$, and satisfies $\lambda_{bs} > 1 + 2z$. (The continuous spectrum is the same as for the impenetrable case.) For $w < 1$, the maximum eigenvalue corresponds to the upper edge of the continuous spectrum and is given by $\lambda_{cont} = 1 + 2z$.

Similar results can be derived for the $3d$ system. Specifically, we have

$$\cosh \varpi = \frac{\lambda - 1 - 2z}{2z}, \tag{4.26}$$

and

$$\lambda - w = 2wz(1 + e^{-\varpi}), \tag{4.27}$$

instead of (4.24)-(4.25). The condition for the existence of the bound state is again $w > w_* \equiv 1$, with $\lambda_{bs} > 1 + 4z$. For $w < 1$ the maximum eigenvalue corresponds to the upper edge of the continuous spectrum: $\lambda_{cont} = 1 + 4z$.

The condition $z\lambda_{bs} = 1$ can be written in the form

$$4w^2 z^4 = (1 - wz)(2w - wz - 1), \tag{4.28}$$

in $2d$, and

$$4w^2 z^4 = (2wz^2 + wz - 1)(2wz^2 + wz - 2w + 1), \tag{4.29}$$

in $3d$. These relations define $z_\infty(w)$ for $w > w_* = 1$. The critical point, $(z_c, w_c)$, is at $(1/2, 1)$ for $2d$ and $((\sqrt{17} - 1)/8, 1)$ for $3d$. (Note that $z_c$ values are identical with those for the impenetrable surface.)

Using the relation (4.21), we find that the fraction of adsorbed monomers $C_s(w) \equiv 0$ for $w \leq 1$. This corresponds to the desorbed regime. Note that $w < 1$ corresponds to a *repulsive* surface, i.e., $\Omega_s < 0$ in (4.1). On the other hand, for $w > 1$ the fraction of adsorbed monomers has a finite value which for $w \gg 1$ can be expanded in the form

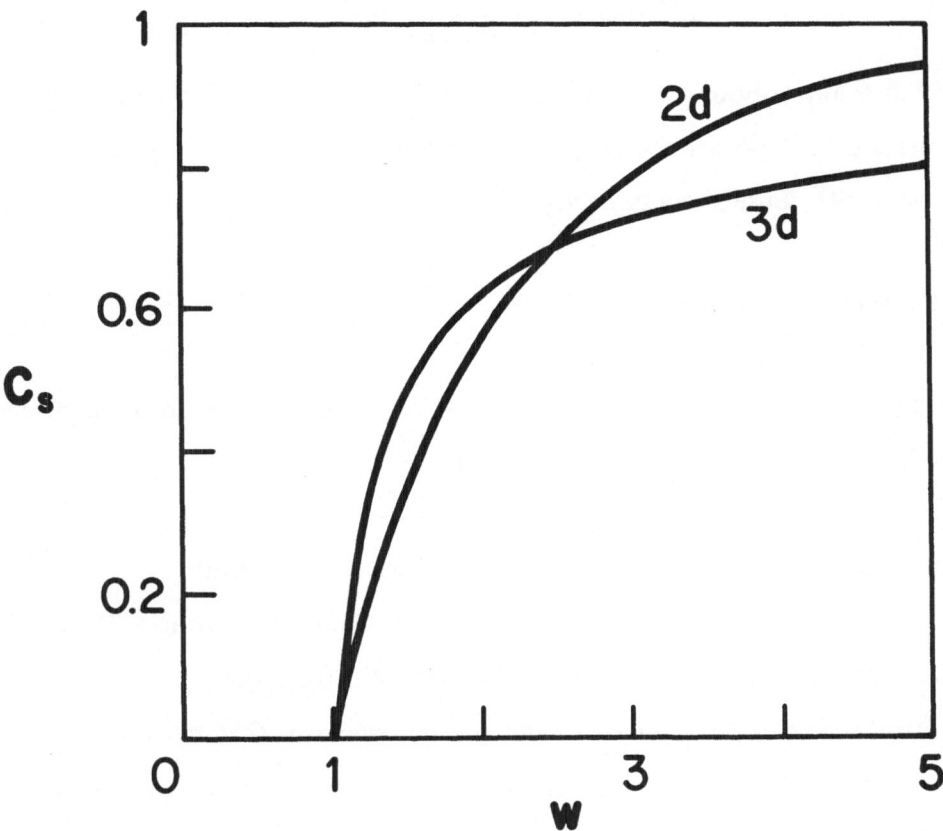

**Figure 7**

Fraction of adsorbed monomers at a symmetric penetrable surface in $2d$ and $3d$.

$$C_s^{(2d)}(w) = 1 - \frac{6}{w^3} - \frac{8}{w^4} - ..., \tag{4.30}$$

and

$$C_s^{(3d)}(w) = 1 - \frac{2}{w} + \frac{12}{w^2} - \frac{86}{w^3} + .... \tag{4.31}$$

The full results, obtained by using (4.21), are shown in Fig. 7. Finally, a detailed calculation (not presented here) of the exponents $\gamma_1$ and $\gamma_{11}$ yields the values identical with those given in Section A.

## C. Selected literature

Theoretical and experimental aspects of the polymer adsorption problem have been reviewed in the following works.

**G.J. Fleer, J.M.H.M. Scheutjens** and **M.A. Cohen Stuart**,

Colloids and Surfaces **31**, 1 (1988)

**P.G. de Gennes**, Adv. Colloid Interface Sci. **27**, 189 (1987)

**K. Binder** and **K. Kremer**, in *Scaling Phenomena in Disordered Systems*,

edited by R. Pynn and A. Skjeltorp (Plenum, NY, 1985), p.525

**A. Takahashi** and **M. Kawaguchi**, Adv. Polymer Sci. **46**, 1 (1982)

**B. Vincent**, Adv. Colloid Interface Sci. **4**, 193 (1974)

Directed walk results presented in this chapter have been collected from the works by V. Privman, G. Forgacs and H.L. Frisch, Phys. Rev. B**37**, 9897 (1988); M.C.T.P. Carvalho and V. Privman, J. Phys A**21**, L1033 (1988). A solvable model of isotropic Gaussian walks at surfaces (not discussed here) has been studied by R.J. Rubin, J. Chem. Phys. **43**, 2392 (1965).

# V. MODELS OF STACKS AND COMPACT CLUSTERS

Cluster models (also termed lattice animal models), together with random walks, surfaces, solid–on–solid strings and sheets, etc., serve as prototype lattice systems with "geometric" phase transitions. Typically, cluster models involve a set of $N$ distinct connected points (sites) or links (bonds), on a $d$–dimensional lattice, with each point connected to the origin through other cluster points; the origin point belongs to the cluster. Geometric connectivity rules, such as compactness, directedness, etc., define different classes of cluster models. A quantity that characterizes statistical "entropic" properties of cluster models is the total number, $c_N$, of *different* $N$-point clusters that can be formed with a prescribed connectivity.

In this chapter we study a class of *compact* $2d$ cluster models which are equivalent to *stacking models*. In many cases they can be solved exactly for the partition functions which generate the cluster numbers $c_N$. Specifically, in Section A, we study the model of stacking of squares at a line wall, which can be easily solved by the generating function method, and serves to illustrate some of the general features of compact cluster models. Section B is devoted to a more complicated circle stacking model. The solution of this problem fully illustrates mathematical complexity of some of the compact cluster models. The methods that are useful in such solutions, namely the generating function and continued fraction techniques, are outlined in Section C. Other stacking models (squares, circles), with different building rules, are described in Section D. Solutions of these models require further mathematical developments, beyond conventional methods, which we outline in Section E. Analysis of finite-size properties of compact lattice animals is presented in Section F, where we also obtain the cluster-radius exponents $\nu_\| = 1$ and $\nu_\perp = 1/2$. Finally, in Section G, we list selected literature.

## A. Stacking of squares at a line wall

Consider a 2d "castle wall" built up from $N$ squares, as shown in Fig. 8. The base row of the cluster must be continuous. Higher rows can have gaps. However, each column must be continuous "self–supporting". Our goal is to calculate the total number $c_N$ of different $N$–site clusters, i.e., the number of possible arrangements of $N$ squares consistent with the above restrictions. In order to avoid counting clusters that differ only by overall translations, we pin the lowest leftmost square with its center at $(X, Y) = (0, 0)$, see Fig. 8. Let $c_{N,k}$ denote the number of different clusters with exactly $k$ squares in the leftmost $X = 0$ column ($k = 2$ in Fig. 8). Obviously, $c_{N,k} = 0$ for $N < k$, and $c_{N,N} = 1$. Define the restricted partition functions which generate the numbers $c_{N,k}$ by

$$F_k(z) = \sum_{N=k}^{\infty} c_{N,k} z^{N-k}, \quad k \geq 1. \tag{5.1}$$

These functions satisfy the following recursion relations,

$$F_k(z) = 1 + \sum_{m=1}^{\infty} z^m F_m(z), \tag{5.2}$$

where each term specifies one possible configuration of the next column centered at $X = 1$. The term 1 corresponds to no second column (the $k = N$ cluster). Each $m > 0$ term in (5.2) sums up all configurations with exactly $m$ squares at $X = 1$, with $F_m(z)$ accounting for all possible arrangements of the remaining squares in the $X > 1$ columns.

The *total* partition function $G(z)$ for the cluster numbers $c_N$, is given by

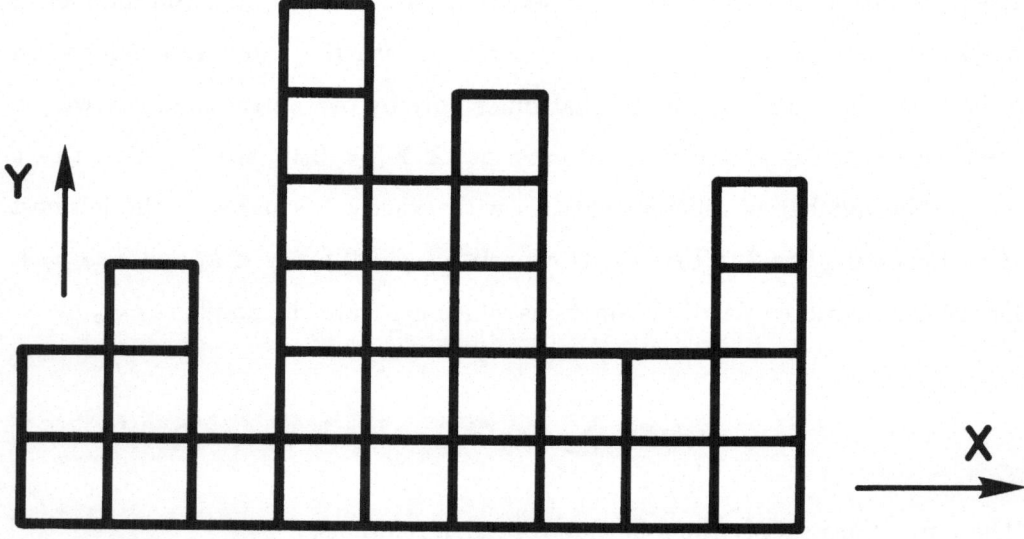

**Figure 8**

Stacking of squares according to the "castle wall" connectivity rules.

$$G(z) \equiv \sum_{N=1}^{\infty} c_N z^N = \sum_{k=1}^{\infty} z^k F_k(z), \tag{5.3}$$

where the last step is self–explanatory [see (5.1)]. Inspection of (5.2) gives

$$F_k(z) = 1 + G(z), \quad \text{for all } k. \tag{5.4}$$

Finally, substitution in (5.2) yields

$$1 + G = 1 + (1 + G) \sum_{m=1}^{\infty} z^m, \tag{5.5}$$

or

$$G(z) = \frac{z}{1 - 2z}, \tag{5.6}$$

The Taylor series coefficients of $G(z)$ are the desired cluster numbers $c_N$ [see (5.3)]. Thus,

$$c_N = 2^{N-1} \tag{5.7}$$

for the square–stacking model.

More generally, the large–$N$ form of $c_N$, applicable to most cluster models is

$$c_N \approx C N^{-\theta} \lambda^N. \tag{5.8}$$

Here $C$ and $\lambda$ are model dependent (nonuniversal), while the "critical exponent" $\theta$ is *universal* for large classes of models which differ by the details of the "microscopic" connectivity rules, lattice structure, etc., but share global "macroscopic" features like directedness, compactness, dimensionality of space. Comparing (5.7) and (5.8) we have $\theta = 0$, for the square–stacking model.

As usual, the large–$N$ behavior of the Taylor coefficients $c_N$, of the partition function is controlled by the singularity nearest to the origin, on the real, positive

$z$ axis (since all $c_N > 0$). In the present case, $G(z)$ is the ratio of two polynomials, and the singularity is a simple pole at $z_c = 1/\lambda = \frac{1}{2}$. [Recall that random walk generating functions (Chapter II) have similar singular behavior.]

## B. Stacking of circles at a line wall

In this section we consider stackings of circles at a line wall as illustrated by the open circles in Fig. 9: $N$ circles are positioned in such a way that the base row is continuous. The higher rows can have gaps, however, each circle must be "supported" by having both lower–$Y$ neighbors occupied. The centers then follow the pattern of the triangular lattice with spacing equal to the circle diameter.

In order to solve this model by the generating function technique, we extend the allowed configurations to include additional $k - 1$ base circles along a lattice direction forming $60°$ with the negative $X$ axis. The case $k = 3$ is illustrated in Fig. 9. The $k - 1 = 2$ full circles are part of the base. Together with the open circles they can "support" additional circles (full circles in Fig. 9).

Let $c_{N,k}$ denote the number of distinct $N$–circle clusters with exactly $k$ circles in the $60°$ base (counting the circle which also belongs to the horizontal base, the length of which is not restricted). The fixed-$k$ partition functions $F_k(z)$ defined as in (5.1), satisfy the recursion relations

$$F_k(z) = 1 + \sum_{m=1}^{k+1} z^m F_m(z), \quad k \geq 1. \tag{5.9}$$

As in Section A, the terms on the right sum up configurations with different number, $m$, of circles in the $60°$ row next to the base $60°$ row. Note that by the stacking rules, $m$ cannot exceed $k + 1$. Replacing (5.9) by the first difference, we get a second–order difference equation

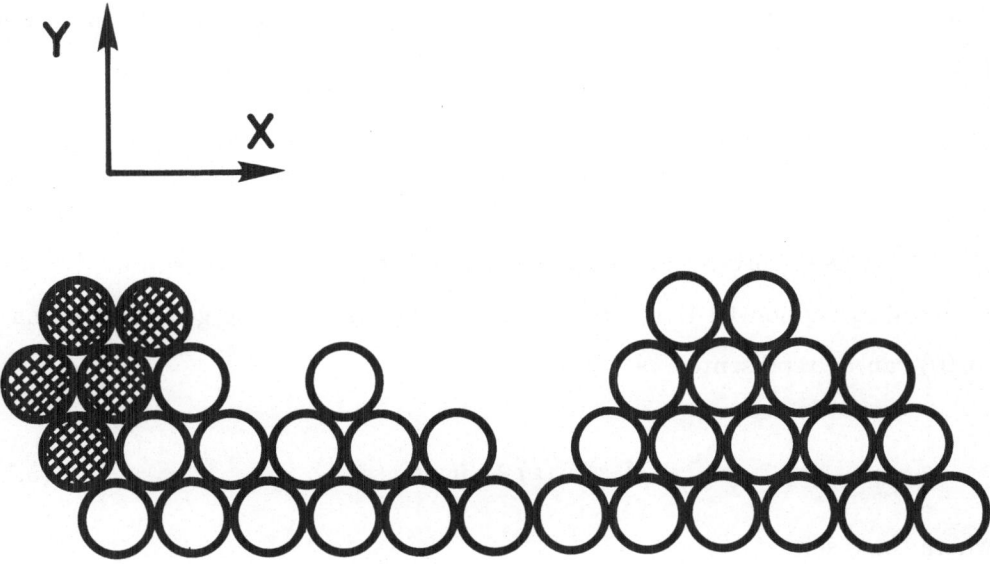

**Figure 9**

Compact self-supporting stacking of circles at a line wall (open circles). Full circles illustrate the two additional 60° base circles, and the circles supported by the extended base.

$$F_{k+1}(z) - F_k(z) = z^{k+2} F_{k+2}(z), \quad k \geq 1, \tag{5.10}$$

with the boundary condition

$$F_1(z) = 1 + z F_1(z) + z^2 F_2(z). \tag{5.11}$$

Note that the partition function for the original circle–stacking problem is given by

$$G(z) = \sum_{N=1}^{\infty} c_{N,1} z^N = z F_1(z). \tag{5.12}$$

Methods of solving difference equations of the type (5.10) will be discussed in detail in Section C. Here, we only quote the results. The general solution of (5.10) can be represented as

$$A(z)\phi_k(z) + B(z)\Phi_k(z) , \tag{5.13}$$

where the $q$-series

$$\phi_k(z) = \sum_{n=0}^{\infty} (-1)^n \frac{z^{n(n+k+1)}}{q_n(z)} , \tag{5.14}$$

with $q_0 \equiv 1$ and

$$q_n(z) \equiv \prod_{j=1}^{n} \left( 1 - z^j \right), \quad \text{for } n \geq 1, \tag{5.15}$$

represents the "physical" or regular at $z = 0$ (for $k \geq -2$) solution. One can show that the second linearly independent solution $\Phi_k(z)$ is power–law–singular at $z = 0$ for sufficiently large $k$. Furthermore, in the mathematical nomenclature, $\phi_k(z)$ is the *minimal* solution in that

$$\lim_{k \to \infty} [\phi_k(z)/\Phi_k(z)] = 0 \tag{5.16}$$

for all $z$ values of "physical" interest, i.e., for $0 < z < 1$. A more detailed discussion of the minimal solution concept is given in Chapter III; see discussion following (3.41).

The boundary condition (5.11) can be satisfied with $B(z) \equiv 0$ and

$$A(z) = [(1 - z)\phi_1(z) - z^2\phi_2(z)]^{-1} \tag{5.17}$$

in (5.13). Thus, the partition function (5.12) for the circle–stacking problem reduces to

$$G(z) = \frac{z}{1 - z - z^2\phi_2(z)/\phi_1(z)}. \tag{5.18}$$

The $q$–series $\phi_k(z)$ are analytic for $|z| < 1$, with a natural boundary of essential singularities $at$ the unit circle. However, the nearest–to–the–origin singularity of $G(z)$ is a simple isolated pole at the first zero of the denominator of (5.18), at $z = z_c = 1/\lambda$, as shown in Fig. 10, similar to other compact cluster models with extensive entropy ($\lambda > 1$). We get

$$z_c = \lambda^{-1} = 0.576148769\ldots; \tag{5.19}$$

furthermore, we have $\theta = 0$ for the universal exponent in (5.8), while for the nonuniversal prefactor we obtain $C = 0.31236\ldots$.

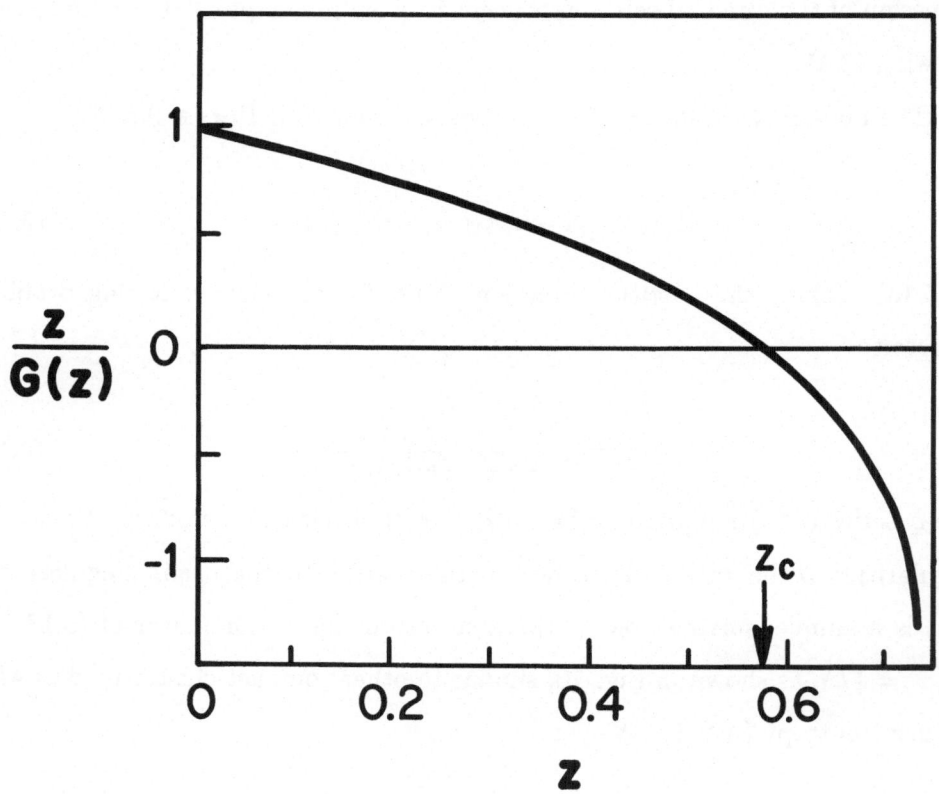

**Figure 10**

The function $z/G(z)$, see equation (5.18). The leading pole in $G(z)$ at $z_c = 0.576...$ is followed by a sequence of zeros-poles (outside the $z$ range shown) which accumulate at $z = 1^-$.

## C. Continued fraction and generating function techniques

In this section we outline two methods which are particularly useful in solution of difference equations associated with stacking models. A standard mathematical approach to linear *second–order* difference equations utilizes *continued fractions* to calculate the minimal solution [see(5.16)]. We illustrate the technique for the circle–stacking problem of Section B.

The difference equation (5.10) is reformulated in terms of the ratios

$$R_k(z) = F_{k+1}(z)/F_k(z) \tag{5.20}$$

as

$$1 - R_k^{-1}(z) = z^{k+2} R_{k+1}, \tag{5.21}$$

$$R_k = \frac{1}{1 - z^{k+2} R_{k+1}}. \tag{5.22}$$

This relation can be iterated to generate a "backward" continued fraction representation

$$R_k(z) = \cfrac{1}{1 - \cfrac{z^{k+2}}{1 - \cfrac{z^{k+3}}{1 - \cfrac{z^{k+4}}{1 - \cdots}}}}. \tag{5.23}$$

The Pincherle theorem (see Chapter III) then ensures a one–to–one correspondence between the convergence of the continued fractions (5.23) and the existence of the minimal solution given by

$$\phi_k(z) = \phi_0(z) \prod_{j=0}^{k-1} R_j(z), \quad k \geq 1. \tag{5.24}$$

(Note that $\phi_k(z)$ is defined up to an arbitrary $z$–dependent coefficient, $\phi_0(z)$, since (5.10) is linear.)

For the "physical" partition function (5.18), one could use the continued fraction representation, e.g., in the form

$$G(z) = \cfrac{1}{1 - \cfrac{z}{1 - \cfrac{z^2}{1 - \cfrac{z^3}{1 - \cdots}}}} - 1, \tag{5.25}$$

involving the so-called Ramanujan's continued fraction. This representation can be used to reproduce all the conclusions on the analytic structure of $G(z)$ mentioned in Section B. The continued fraction representation is also not inferior to the infinite–sum forms like (5.14), for numerical computation purposes. However, the infinite–sum representations are more familiar. They can be obtained in many cases by utilizing sum-product identities from the classical works of Ramanujan.

For physical applications and in particular, to analyze the complex–plane singularities, an *infinite–product* representation of a partition function would be valuable. We are not aware of any mathematical results appropriate for $G(z)$ here. However, infinite–product forms have been utilized in other applications of the $q$–series in Physics. We will see in Section D, that in some cases both the minimal and some of the other linearly independent solutions are physically relevant. When one solution (minimal) is available, the order of the difference equation can be reduced by one, by standard methods.

The *generating function approach* to linear second-order equations consists of considering the generating function

$$P(z,t) = \sum_{k=1}^{\infty} F_k(z)t^{k-1}. \tag{5.26}$$

When the difference equation is multiplied by $t^{k-1}$ and summed over $k$, one sometimes ends up with a tractable equation for $P(z,t)$. Specifically, for difference equations with constant ($k$–independent) coefficients, algebraic equations for $P$ are obtained. Equations with coefficients involving integral powers of $k$ yield *differential equations* for $P$; this case is important in the solid–on–solid model studies (see Chapter III). However, difference equations with exponential–in–$k$ coefficients, like $z^k$, lead to functional equations for $P(z,t)$ which are rather difficult to solve in general. The generating function method, and related techniques, e.g., Laplace's method, complement the continued fraction approach. However, they can also be used for equations of order higher than second, and in some cases are advantageous even for second–order difference equations.

For compact cluster models considered here, one typically obtains a functional equation for $P(z,t)$, of the form

$$P(z,t) = a(z,t) + b(z,t)P(z,tz). \tag{5.27}$$

Mathematical literature on equations of this sort is limited. When the resulting series is well defined, one can use a solution obtained by iterating (5.27) an infinite number of times,

$$P(z,t) = a(z,t) + \sum_{n=1}^{\infty} \left[ a\left(z,tz^n\right) \prod_{m=0}^{n-1} b\left(z,tz^m\right) \right]. \tag{5.28}$$

The solution of (5.27) is linear in $a$ in the sense that for

$$a(z,t) = \alpha_1(z)a_1(z,t) + \alpha_2(z)a_2(z,t), \tag{5.29}$$

one has

$$P_a = \alpha_1(z)P_{a_1} + \alpha_2(z)P_{a_2}, \tag{5.30}$$

as can be seen explicitly for (5.28).

For the circle stacking problem considered in Section B, the appropriate equation (5.27) has coefficients

$$a(z,t) = \frac{F_1(z) - z^2 F_2(z)}{1-t} - \frac{zF_1(z)}{t(1-t)}, \tag{5.31}$$

$$b(z,t) = \frac{z}{t(1-t)}. \tag{5.32}$$

Each solution is a linear combination of the type (5.30), with $a_1 = (1-t)^{-1}$, $a_2 = [t(1-t)]^{-1}$. The coefficients $\alpha_{1,2}(z)$ involve two unknown functions $F_{1,2}(z)$. One relation between the coefficients is provided by the boundary condition (5.11). Another condition must therefore result from the "analyticity" requirement on $P(z,t)$, near $t = 0$.

However, expansion (5.28) is ill–defined in this case. The generating function method can be applied to this problem if we consider modified functions

$$\mathcal{F}_k(z) = z^{k(k+1)/2} F_k(z) \tag{5.33}$$

satisfying

$$\mathcal{F}_{k+1}(z) - \mathcal{F}_{k+2}(z) = z^{k+1} \mathcal{F}_k(z). \tag{5.34}$$

After a tedious calculation utilizing (5.28), one ends up with an expression for the partition function $G(z)$. However, it is much more complicated than (5.18) or (5.25). We omit the details of this calculation.

## D. Other stacking models

Using the techniques outlined in the previous section we can obtain analytic results for the generating functions of a variety of cluster models with different stacking rules. Several of them are described in this section.

Consider first a *"one–tooth" stacking of squares* in a pyramid–like shape illustrated in Fig. 11: The model is similar to that of Section A, but with the additional requirement that each row is continuous. The quantity of interest is the total number $c_N$ of pyramids that can be built from $N$ squares. Let $c_{N,k}$ denote the number of distinct $N$–square pyramids with exactly $k$ squares in the base row. As usual, we define the restricted partition functions by (5.1). The appropriate recursion relations are

$$F_k = 1 + \sum_{m=1}^{k}(k - m + 1)z^m F_m. \tag{5.35}$$

The important change here, as compared to the recursions considered in Sections A-B, is the factor $(k - m + 1)$ accounting for the number of ways in which the second horizontal row of $m \leq k$ squares can be positioned. The $k = 1, 2$ relations provide the boundary conditions which can be simplified to

$$F_1(z) = \frac{1}{1 - z}, \quad F_2(z) = \frac{1}{(1 - z)^2}. \tag{5.36}$$

For higher $k$ relations, we form second differences to obtain

$$F_{k+2} - 2F_{k+1} + F_k = z^{k+2}F_{k+2}. \tag{5.37}$$

Examination of (5.35) suggests that all $F_k$ are rational functions of the form

$$F_k(z) = Q_k(z)/q_k(z) \tag{5.38}$$

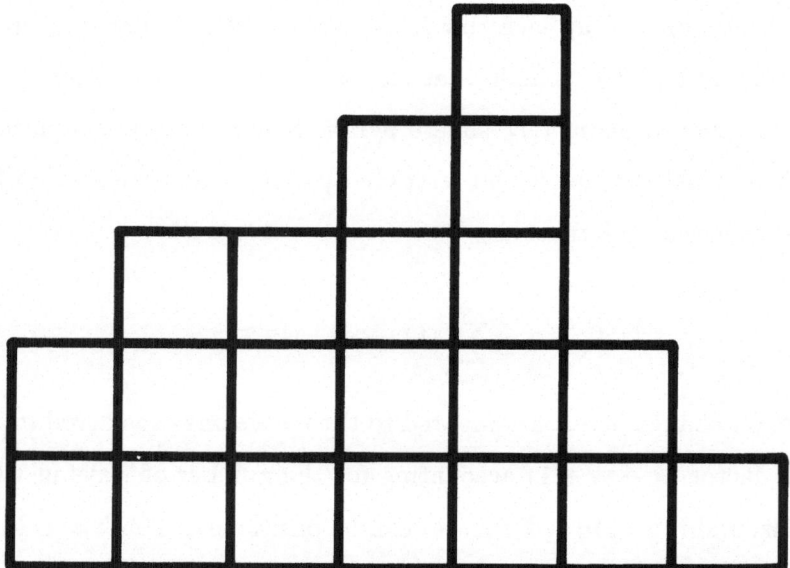

**Figure 11**

Stacking of squares in a pyramid-like shape.

[see (5.15)], where $Q_k$ are polynomials. Finally, note that the total partition function for the single–pyramid problem is given by

$$G(z) = P(z, z), \tag{5.39}$$

[see (5.3), (5.26)].

Let us apply the generating function method to (5.36) – (5.37). Proceeding along the lines of Section C, we get

$$P(z, t) = \frac{(1 - z - 2t)F_1(z) + t(1 - z^2)F_2(z)}{(1 - t)^2} + \frac{z}{(1 - t)^2} P(z, tz). \tag{5.40}$$

Since there are two boundary conditions, we expect that *both* linearly independent solutions are physically acceptable. In this case $F_{1,2}$ are known explicitly. Thus, (5.40) reduces to (5.27) with

$$a = \frac{1}{1 - t}, \qquad b = \frac{z}{(1 - t)^2}. \tag{5.41}$$

By using (5.28), $P(z, t)$ is obtained as

$$P(z, t) = \sum_{n=0}^{\infty} \frac{z^n(1 - tz^n)}{\left[ \prod_{m=0}^{n} (1 - tz^m) \right]^2}. \tag{5.42}$$

For $t = z$, this is identical with the result obtained by Temperley for $G(z)$ [see (5.39)] by a different method,

$$G(z) = \sum_{n=0}^{\infty} \frac{z^n(1 - z^{n+1})}{q_{n+1}^2}. \tag{5.43}$$

This function has an essential singularity at $z = 1$, analysis of which yields the large–$N$ cluster numbers as

$$c_N \approx C N^{-5/4} \mu^{\sqrt{N}}, \qquad (5.44)$$

where $C$ and $\mu$ are known constants. This result is different from the generic lattice animal form (5.8). It is interesting to mention in this connection a model of filling a corner by squares obtained by imposing additional constraint in the one–pyramid square stacking model, namely that the leftmost column must not be shorter than any other column in the cluster. The resulting model is equivalent to enumeration of nonincreasing partitions of $N$. Thus, detailed results are available, specifically,

$$c_N \approx C N^{-1} \mu^{\sqrt{N}} \qquad (5.45)$$

(with different $C$ and $\mu$). The exponent of the power law prefactors in (5.44) - (5.45) is not universal, unlike $\theta$ in (5.8).

Turning back to the one–pyramid packings of squares, we derive an explicit form for the restricted partition functions $F_k(z)$ by expanding (5.42) in powers of $t$. We use the identity

$$\left[\prod_{m=0}^{n-1}(1 - tz^m)\right]^{-1} = \sum_{j=0}^{\infty} \frac{q_{n+j-1}(z)t^j}{q_{n-1}(z)q_j(z)}, \qquad (5.46)$$

to get

$$F_k(z) = \sum_{n=0}^{\infty} \frac{z^n}{q_{n-1}^2} \left(\tau_{k,n} - z^n \tau_{k-1,n}\right), \qquad (5.47)$$

where

$$\tau_{k,n} \equiv \sum_{j=0}^{k} \frac{q_{n+j-1}(z)q_{n+k-j-1}(z)}{q_j(z)q_{k-j}(z)}. \qquad (5.48)$$

This double–sum representation is rather complicated. (See further below.)

The continued fraction method can be used to derive the form of the minimal solution of (5.37). We will only quote some results here. Continued fractions of the type appropriate for (5.37), have been analyzed by Ramanujan. As a result, an infinite series representation is available,

$$\phi_k(z) = \sum_{n=0}^{\infty} \frac{z^{n(n+k+1)}}{q_n^2}.$$ (5.49)

However, for this problem the other linearly independent solution, $\Phi_k$, is also physically admissible. By reducing the order of the equation using a known solution, we get an extremely complicated result.

However, a relatively simple representation for $\Phi_k(z)$ can be obtained by the method described later in Section E. Thus,

$$\Phi_k(z) = k + \sum_{n=1}^{\infty} \frac{z^{n(n+k+1)}}{q_n^2(z)} \left[k + s_n(z)\right],$$ (5.50)

where

$$s_n(z) \equiv 2 \sum_{j=1}^{n} \frac{1}{1 - z^j}, \qquad \text{for} \quad n \geq 1.$$ (5.51)

By imposing the boundary conditions, we get

$$F_k = \frac{(1-z)(\phi_k \Phi_2 - \phi_2 \Phi_k) - (\phi_k \Phi_1 - \phi_1 \Phi_k)}{(1-z)^2 (\phi_1 \Phi_2 - \phi_2 \Phi_1)}.$$ (5.52)

Another solution of (5.37), $\psi_k(z)$, linearly independent of $\phi_k(z)$, is known in the theory of $q$–ultraspherical polynomials,

$$\psi_k(z) = \sum_{m=0}^{k} \frac{1}{q_m q_{k-m}}.$$ (5.53)

This can be used in place of $\Phi_k$ in (5.52).

Pyramid–shape or *"one tooth" stacking of circles* is illustrated in Fig. 12. The stacking rules are identical to those of Section B (Fig. 9), but now each row must be continuous (no gaps). Let $c_{N,k}$ denote the number of distinct $N$–circle clusters with $k$ circles in the base. Note that $c_N \equiv \sum_{k=K}^{N} c_{N,k}$, where $K$ is the smallest integer greater or equal $(\sqrt{8N+1} - 1)/2$. The restricted generating functions satisfy

$$F_k(z) = 1 + \sum_{m=1}^{k-1} (k - m) z^m F_m(z), \qquad k \geq 2, \tag{5.54}$$

with $F_1(z) = 1$. By forming the second difference, this is reduced to

$$F_{k+2} - 2F_{k+1} + F_k = z^{k+1} F_{k+1}, \qquad k \geq 1, \tag{5.55}$$

with the boundary conditions

$$F_1 = 1, \qquad F_2 = 1 + z. \tag{5.56}$$

Examination of (5.54) leads to the conclusion that all $F_k(z)$ are polynomials of degree $k(k-1)/2$. The structure of this problem is quite similar to that of the pyramid–of–squares stackings considered above. Therefore, we only list some central results here.

The generating function method can be invoked for the pyramid circle stackings. Similarly to the pyramid-of-squares, the boundary conditions (5.56) are used to yield

$$a(z,t) = \frac{1}{1-t}, \qquad b(z,t) = \frac{zt}{(1-t)^2}. \tag{5.57}$$

Thus, (5.28) takes the form

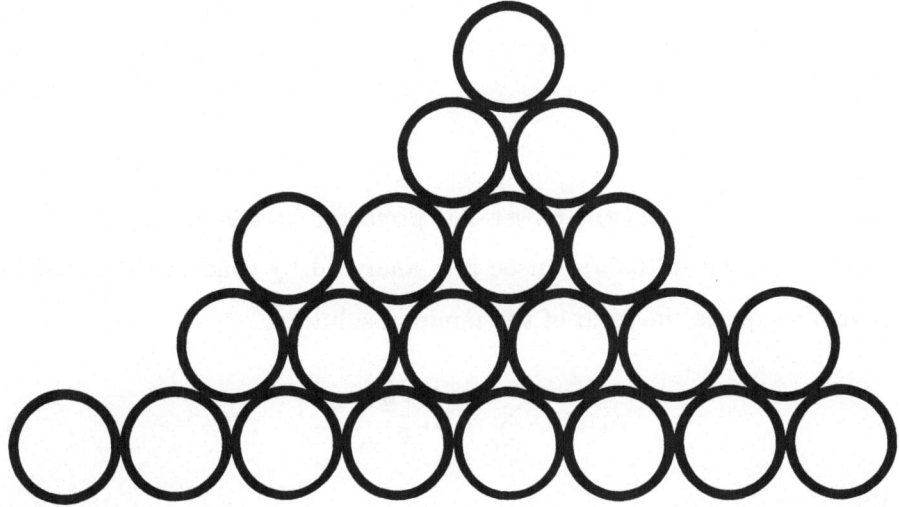

**Figure 12**

Pyramid-shape stacking of circles at a line wall.

$$P(z,t) = \sum_{n=0}^{\infty} \frac{t^n z^{n(n+1)/2}(1 - tz^n)}{\left[\prod_{m=0}^{n}(1 - tz^m)\right]^2}. \tag{5.58}$$

The total partition function, first obtained by Auluck by a different method, is given by

$$G(z) = P(z,z) = \sum_{n=1}^{\infty} \frac{z^{(n-1)(n+2)/2}(1 - z^n)}{q_n^2(z)}. \tag{5.59}$$

This function has an essential singularity at $z_c = 1$, analysis of which yields

$$c_N \sim \mu^{\sqrt{N}}. \tag{5.60}$$

(Unfortunately, the form of the power–law prefactor here is not known.)

Difference equation (5.55) can be also analyzed by continued fraction techniques: we only quote the form of the minimal solution,

$$\phi_k(z) = \sum_{n=0}^{\infty} \frac{z^{n(n+2k+1)/2}}{q_n^2(z)}. \tag{5.61}$$

This infinite series form of $\phi_k(z)$ follows from yet another of the Ramanujan's results. In fact, both solutions of (5.55) are physically acceptable. The second solution can be found by the method of Section E,

$$\Phi_k(z) = k + \sum_{n=1}^{\infty} \frac{z^{n(n+2k+1)/2}}{q_n^2}(k - n + s_n). \tag{5.62}$$

The boundary conditions then imply

$$F_k = \frac{(1 + z)(\phi_1 \Phi_k - \phi_k \Phi_1) - (\phi_2 \Phi_k - \phi_k \Phi_2)}{\phi_1 \Phi_2 - \phi_2 \Phi_1}. \tag{5.63}$$

Next, let us consider the *partially directed compact lattice animal* model, illustrated in Fig. 13. The model is most easily described as having $N$ squares

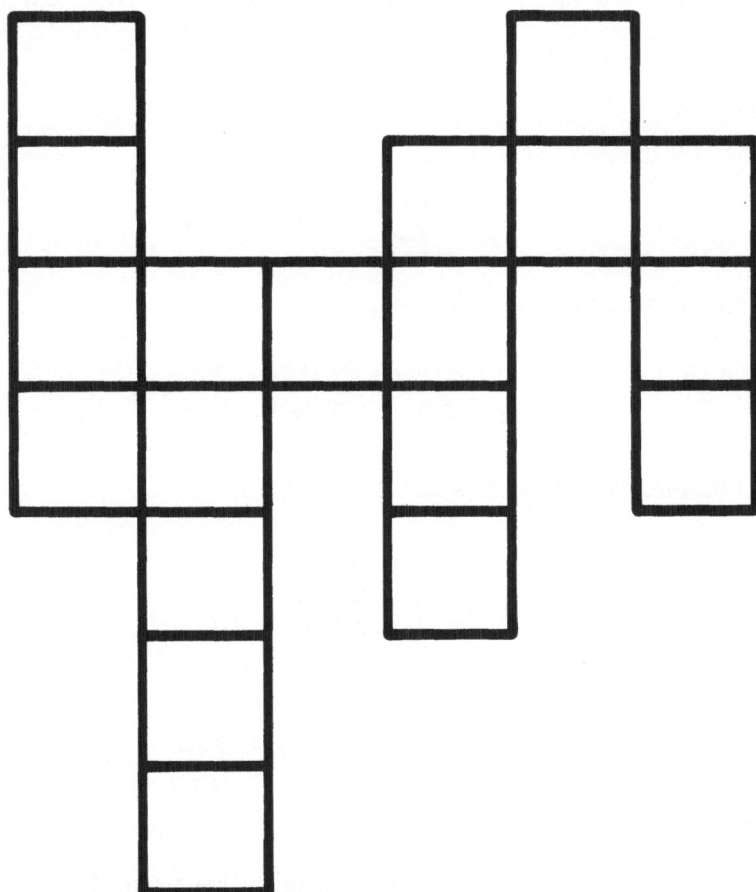

**Figure 13**

A partially directed compact lattice animal.

positioned in (continuous) columns. The neighboring columns must touch by at least one square. A formulation according to directed square lattice animal rules is also possible.

Let $k$ denote the number of squares in the leftmost "root" column, and $c_{N,k}$ be the number of distinct $N$–square $k$–root clusters. Then the restricted partition functions (5.1) satisfy

$$F_k(z) = 1 + \sum_{m=1}^{\infty} (k+m-1)\, z^m F_m(z). \tag{5.64}$$

By forming the second difference, we get

$$F_{k+2} - 2F_{k+1} + F_k = 0, \quad k \geq 1, \tag{5.65}$$

with the boundary conditions ($k = 1, 2$ in (5.64)),

$$F_1 = 1 + \sum_{m=1}^{\infty} m z^m F_m, \tag{5.66}$$

$$F_2 = 1 + \sum_{m=1}^{\infty} (m+1) z^m F_m. \tag{5.67}$$

This problem is interesting in that (5.65) has constant coefficients. Thus, it can be solved in full detail, including for finite-size properties: see Section F. Specifically, the minimal solution is simply $\phi_k(z) = 1$, while the other linear independent solution is $\Phi_k(z) = k$. Both solutions are physically acceptable. We have

$$F_k(z) = k A(z) + B(z). \tag{5.68}$$

The coefficient functions $A$ and $B$ are determined by (5.66)-(5.67). One gets

$$F_k(z) = \frac{kz(1-z)^3 + (1-3z+z^2)(1-z)^2}{1-5z+7z^2-4z^3}. \tag{5.69}$$

The total partition function is given by

$$G(z) = \sum_{k=1}^{\infty} z^k F_k(z) = \frac{z(1-z)^3}{1-5z+7z^2-4z^3}. \tag{5.70}$$

All the partition functions, (5.69) - (5.70), have a simple pole singularity at $z_c = \lambda^{-1}$, where

$$\lambda = 3.20556943\ldots \tag{5.71}$$

or

$$\lambda = \frac{12}{(6\sqrt{177}-71)^{1/3} - (6\sqrt{177}+71)^{1/3} + 7}. \tag{5.72}$$

Thus, for this model (5.8) applies, with $\theta = 0$.

Finally, consider the *fully directed compact lattice animal* model. Although this model can be defined according to the square lattice directed animal rules, it can also be described as stackings of circles: see Fig. 14. Continuous (no horizontal gaps) rows of circles are put on top of each other with the requirement that each circle not in the base is supported by having at least *one* of its lower neighbors present. (Thus, the difference with the pyramid stackings of circles is that there "support" meant both lower neighbors occupied.) The whole cluster is supported by a single base circle (*origin* in the lattice animal formulation).

As usual, we consider the restricted partition functions $F_k(z)$ for the numbers $c_{N,k}$ of $N$–circle, $k$–base clusters, satisfying

$$F_k(z) = 1 + \sum_{m=1}^{k+1} (k-m+2)z^m F_m(z). \tag{5.73}$$

**Figure 14**

A fully directed compact lattice animal.

Except for the two boundary conditions,

$$F_1 = 1 + 2zF_1 + z^2 F_2, \tag{5.74}$$

$$F_2 = 1 + 3zF_1 + 2z^2 F_2 + z^3 F_3, \tag{5.75}$$

the recursions (5.73) can be reduced to

$$F_{k+2}(z) - 2F_{k+1}(z) + F_k(z) = z^{k+3} F_{k+3}(z). \tag{5.76}$$

This is a third order difference equation. Since there are two boundary conditions, one anticipates two physically "regular" and one "irregular" solutions. Various studies have found a simple pole singularity in the partition functions (i.e., $\theta = 0$ in (5.8)), at $z_c = \lambda^{-1}$ with

$$\lambda = 2.661857944\dots . \tag{5.77}$$

The exact solution of this model has been achieved along the following lines. Note that all the difference equations encountered in the stacking models have discrete first or second derivatives on their left hand sides. This is related to the fact that the multiplicity factors, like $(k - m + 2)$ in (5.73) are, respectively, constants or linear functions, in their $k$–dependence. By inspecting the minimal solutions obtained for the second–order difference equations in the preceding examples, we can *guess* one solution for

$$v_{k+1}(z) - v_k(z) = z^{k+m} v_{k+l}(z), \quad l \geq 0. \tag{5.78}$$

It is

$$v_k(z) = \sum_{n=0}^{\infty} \frac{(-1)^n}{q_n} z^{n[l(n-1)+2(k+m)]/2}. \tag{5.79}$$

Similarly,

$$u_{k+2}(z) - 2u_{k+1}(z) + u_k(z) = z^{k+m} u_{k+l}(z), \quad l \geq 0, \tag{5.80}$$

is solved by

$$u_k(z) = \sum_{n=0}^{\infty} q_n^{-2} z^{n[l(n-1)+2(k+m)]/2}. \tag{5.81}$$

Thus, we have one solution for (5.76),

$$\phi_k(z) = \sum_{n=0}^{\infty} \frac{z^{n(3n+2k+3)/2}}{q_n^2}, \tag{5.82}$$

regular at $z = 0$ for $k \geq -3$. Formally, one can then reduce the order of the difference equation and apply the continued fraction method to derive the second regular solution. However, the resulting expressions turn out to be extremely complicated. The generating function method is also not useful because (5.28) is ill-defined.

The second "physical" solution, $\Phi_k(z)$, has been obtained by a different method, described in the next section. We only quote the result for $\Phi_k$ here

$$\Phi_k = k + \sum_{n=1}^{\infty} \frac{z^{n(3n+2k+3)/2}}{q_n^2} (k+n+s_n). \tag{5.83}$$

Thus, for the original problem (5.76) we have

$$F_k(z) = A(z)\phi_k(z) + B(z)\Phi_k(z), \tag{5.84}$$

with $A(z)$ and $B(z)$ determined by the boundary conditions (5.74)-(5.75). Then the total partition function is given by $G(z) = zF_1(z)$. After a straightforward but long algebra we obtain

$$G(z) = z\frac{(1 - z^2)O_{12}(z) - z^3 O_{13}(z)}{(1 + z)(1 - 3z + z^2)O_{12}(z) - z^3(1 - 2z)O_{13}(z) + z^5 O_{23}(z)}, \quad (5.85)$$

where

$$O_{ij}(z) = \phi_i(z)\Phi_j(z) - \phi_j(z)\Phi_i(z). \quad (5.86)$$

As in the case of several other stacking problems considered in this section, the generating function $G(z)$ has intricate pattern of singularities in the complex-$z$ plane. (The behavior of this function is similar to the circle stacking problem, Section B, as shown in Fig. 10.) However, for models with $\lambda > 1$, the leading large-$N$ growth rate is determined by the singularity (simple pole) nearest to the origin, at $z_c = 1/\lambda$.

The solutions of several stacking models presented here illustrate the important features of this class of *compact* animals. If the stacking rules are sufficiently relaxed to allow clusters with finite entropy $N^{-1}\ln c_N \to \ln \lambda$ per element (circle, square) in the large–$N$ limit, i.e., $\lambda > 1$, then the universal form (5.8) applies with $\theta = 0$. Models with more restrictive rules have entropy vanishing as $N^{-1/2}$ [see (5.44), (5.45), (5.60)].

## E. Asymptotic properties of the physical solutions

In this section we develop some mathematical aspects of the solution of stacking models considered above. The nonautonomous difference equations encountered in several compact cluster models considered in this chapter, were always of the form (5.78) or (5.80). Depending on the value of the nonnegative integer $l$, these equations may have several solutions. However, the physically acceptable solutions must be regular for small $z$. Furthermore, the series

$$\sum_{k=1}^{\infty} z^k F_k(z) \tag{5.87}$$

in (5.3), etc., must converge in the physically relevant part of the range $0 \le z < 1$ since (5.87) always represents some sort of a partition function.

Thus, for large $k$, the right sides of (5.78) and (5.80) asymptotically vanish for the "physical" solutions. It follows that there is exactly one such solution, (5.79), of (5.78), i.e., $\phi_k(z) \equiv v_k(z)$. It has a finite limit $v_\infty(z) \equiv 1$ as $k \to \infty$ for fixed $0 \le z < 1$. Difference equation (5.80), however, has *two* "physical" solutions, $\phi_k(z) \equiv u_k(z)$ of (5.81), and $\Phi_k(z)$, with large–$k$ behaviors

$$\phi_k(z) \to 1, \tag{5.88}$$

and

$$\Phi_k(z) \sim \rho_\infty(z)k, \quad \text{for } 0 \le z < 1. \tag{5.89}$$

It is interesting to recall that we always had exactly the right number of boundary conditions, (5.11), (5.36), (5.56), (5.74)-(5.75), requiring one solution of (5.78), but two solutions of (5.80).

The asymptotically constant solution, which tends to unity when $k \to \infty$, is obtained from the following ansatze: For (5.78), we try

$$\phi_k(z) = \sum_{n=0}^{\infty} q_n^{-1} z^{kn} f_n(z), \tag{5.90}$$

which, upon substitution into (5.78), yields a *first-order* difference equation for $f_n(z)$, which can then be solved by a straightforward iteration. Similarly, for (5.80), the appropriate ansatz is

$$\phi_k(z) = \sum_{n=0}^{\infty} q_n^{-2} z^{kn} f_n(z), \tag{5.91}$$

which also leads to a first-order equation for $f_n(z)$. (A similar approach was used in our solution of the difference equation (3.44) in Chapter III.)

The second physical solution of (5.80), asymptotically linear in $k$, can be obtained in the form

$$\Phi_k(z) = k\phi_k(z) + g_k(z). \tag{5.92}$$

Substitution in (5.80) and use of the fact that $\phi_k$ is a solution, yield after some algebra the following inhomogeneous equation for $g_k(z)$,

$$g_{k+2} - 2g_{k+1} + g_k = z^{k+m} g_{k+l} + (l-2)\phi_{k+2} - 2(l-1)\phi_{k+1} + l\phi_k. \tag{5.93}$$

This equation can be solved by the ansatz

$$g_k = \sum_{n=0}^{\infty} \frac{z^{n[l(n-1)+2(k+m)]/2}}{q_n^2(z)} p_n(z), \tag{5.94}$$

inspired by (5.91). Indeed, after a long but straightforward calculation, one concludes that $p_n$ must satisfy the first-order equation

$$p_{n+1}(z) = p_n(z) + l - 2 + \frac{2}{1 - z^{n+1}}, \qquad \text{for} \quad n \geq 0. \tag{5.95}$$

Note that $\Phi_k(z)$ can be redefined up to an additive term of the form $h(z)\phi_k(z)$. This allows the convenient choice $p_0(z) \equiv 0$, yielding

$$p_n(z) = n(l-2) + s_n(z), \quad \text{for} \ n \geq 1, \tag{5.96}$$

where $s_n(z)$ is defined by (5.51). Finally, the second "physical" solution of (5.80) is obtained as

$$\Phi_k(z) = k + \sum_{n=1}^{\infty} \frac{z^{n[l(n-1)+2(k+m)]/2}}{q_n^2} \left[ k + n\left(l-2\right) + s_n \right]. \tag{5.97}$$

Note that this choice corresponds to $\rho_\infty(z) \equiv 1$ in (5.89). With the appropriate values of $l$ and $m$ in (5.97) we obtain the relations (5.50), (5.62) and (5.83).

## F. Finite-size and growth properties of compact animals

In order to analyze the finite-size properties of compact clusters, we consider the partially directed compact lattice animal model, described in Section C and Fig. 13. This model is of particular interest because some of its finite-size properties can be obtained analytically. It is therefore instructive to present the calculation in some detail.

Let us first consider the definition of dimensions of directed lattice animals. For the $2d$ partially directed compact lattice animals of size $N$, one can define (see below) two cluster radii, $R_\parallel(N)$ and $R_\perp(N)$, which grow according to

$$R_\parallel(N) \sim N^{\nu_\parallel} \quad \text{and} \quad R_\perp(N) \sim N^{\nu_\perp}, \tag{5.98}$$

as $N \to \infty$ [compare (2.8)-(2.9)]. The indices $\parallel$ and $\perp$ indicate cluster dimensions parallel and orthogonal to the directed axis (horizontal, $X$ axis).

Consider the $N$-site partially directed compact lattice animal as shown in Fig. 13, with the leftmost "root" column (of $k$ sites) positioned at $X = 0$.

Denote by $x_n$ the $X$ coordinates of the sites: $n = 1, 2, ..., N$. Let the index $a$ label all $k$-root animals, and define the generating function

$$F_k(z, u) = \sum_a z^{N(a)-k} u^{\Xi(a)}, \quad \Xi(a) \equiv \sum_{n=1}^{N(a)} x_n(a), \qquad (5.99)$$

where $N(a)$ is the number of sites in the $a$th animal and $x_n(a)$ are the appropriate $X$ coordinates. This definition reduces to (5.1) for $u \equiv 1$, i.e., $F_k(z, 1) = F_k(z)$, satisfying (5.64). A measure of animal dimension along the $X$ axis can be defined by averaging the center-of-mass $X$ coordinate

$$\frac{1}{N} \sum_{n=1}^{N} x_n = \frac{\Xi}{N} \qquad (5.100)$$

over the different $N$-site animals. Summing over *all* the $N$-site animals (labeled by index $b$), we have

$$R_{\|}(N) = \frac{1}{c_N} \sum_{b=1}^{c_N} \left[ N^{-1} \sum_{n=1}^{N} x_n(b) \right], \qquad (5.101)$$

where $c_N$ is the total number of different $N$-site animals. [Note that first-moment definition for $R_\perp(N)$ vanishes identically.] By (5.8) and (5.98),

$$N c_N R_{\|}(N) \equiv \sum_{b=1}^{c_N} \sum_{n=1}^{N} x_n(b) \sim N^{\nu_{\|} - \theta + 1} \lambda^N. \qquad (5.102)$$

It is convenient to define $R_{\|}(N, k)$ for $k$-root clusters only. The relevant generating function takes the form

$$R_k(z) \equiv \left[ \frac{\partial F_k(z, u)}{\partial u} \right]_{u=1} = \sum_{N=k}^{\infty} N c_N R_{\|}(N, k) z^{N-k}. \qquad (5.103)$$

After these preliminary definitions we now calculate $F_k(z, u)$. Note that these generating functions satisfy

$$F_k(z, u) = 1 + \sum_{m=1}^{\infty} (k + m - 1) \sum_a z^{N(a)} u^{\Xi(a)+N(a)}, \tag{5.104}$$

which reduces to (5.64) for $u \equiv 1$. This can be further rewritten as

$$F_k(z, u) = 1 + \sum_{m=1}^{\infty} (k + m - 1)(zu)^m F_m(zu, u). \tag{5.105}$$

The $u \equiv 1$ relation is solved by (5.69), giving $c_N$ in the form (5.8), with $\theta = 0$ and $\lambda > 1$ [see (5.71)]. In order to obtain $R_\parallel(N)$ we calculate the derivative as indicated in (5.103). Using (5.105) with $z$ replaced by $z/u$, we get

$$R_k(z) = z \frac{dF_k(z, 1)}{dz} + \sum_{m=1}^{\infty} (k + m - 1) z^m R_m(z). \tag{5.106}$$

By forming the second difference and using (5.65), this can be further rewritten in the form

$$R_{k+2}(z) - 2R_{k+1}(z) + R_k(z) = 0, \tag{5.107}$$

giving

$$R_k(z) = kC(z) + D(z), \tag{5.108}$$

where $C(z)$ and $D(z)$ are determined from the boundary condition obtained from (5.106) with $k = 1, 2$. After a long algebra one gets

$$C(z) = \frac{z(1 - z)^4(1 - 2z)(1 - 4z + 10z^2 - 8z^3 + 8z^4 - 2z^5)}{(1 - 5z + 7z^2 - 4z^3)^3}, \tag{5.109}$$

$$D(z) = \frac{2z^2(1 - z)^3(1 - 5z + 10z^2 - 12z^3 + 15z^4 - 7z^5 + 2z^6)}{(1 - 5z + 7z^2 - 4z^3)^3}. \tag{5.110}$$

By using (5.104), the generating function for the quantities $Nc_N R_\parallel(N)$ can be expressed in terms of $R_k(z)$ as

$$E(z) \equiv \sum_{N=1}^{\infty} [Nc_N R_\parallel(z)] z^N = \sum_{k=1}^{\infty} z^k R_k(z), \qquad (5.111)$$

which, by (5.108), gives

$$E(z) = z(1-z)^{-2} [C(z) + (1-z)D(z)]. \qquad (5.112)$$

The asymptotic behavior of (5.102) corresponds to the $(1-\lambda z)^{-(\nu_\parallel + 2 - \theta)}$ singularity of $E(z)$. From (5.109), (5.110) and (5.112), it follows that $\nu_\parallel = 1$ for compact directed animals. The corresponding calculation of $\nu_\perp$, characterizing the cluster dimension along the symmetric $(Y)$ axis, requires the use of the second or higher even-power moments of the size distribution. This calculation (not detailed here) has been performed on a computer, with the result $\nu_\perp = 1/2$. (Recall that similar exponent values have been obtained for the PDSAW in Chapter II.)

The caliper size distribution along the $X$ axis can be also calculated. Let $c_{N,k}(L)$ denote the number of distinct $N$-site $k$-root animals with exactly $L$ columns, i.e., with $X$ ranging from 0 to $L-1$. Obviously, we have the relations

$$c_{N,k}(L) = \begin{cases} 0, & \text{for } N < k + (L-1), \\ k, & \text{for } N = k + (L-1), \\ \delta_{N,k}, & \text{for } L = 1. \end{cases} \qquad (5.113)$$

Recursion relations for $c_{N,k}(L)$ are

$$c_{N,k}(L) = \sum_{m=1}^{N-k-(L-2)} (k+m-1)c_{N-k,m}(L-1), \qquad (5.114)$$

where the upper limit for $m$ is obtained from the condition $m_{max} + (L-2) = N-k$. It is convenient to introduce a double-generating function,

$$G_k(z,v) = \sum_{L=1}^{\infty} \sum_{N=k+(L-1)}^{\infty} c_{N,k}(L) z^{N-k} v^L, \tag{5.115}$$

which, with (5.113), reduces to

$$G_k(z,v) - v = \sum_{L=2}^{\infty} \sum_{N=k+(L-1)}^{\infty} c_{N,k}(L) z^{N-k} v^L. \tag{5.116}$$

By (5.114), we obtain a triple sum which can be rearranged to give

$$G_k(z,v) = v \left[ 1 + \sum_{m=1}^{\infty} (k+m-1) z^m G_m(z,v) \right]. \tag{5.117}$$

Form this relation $G_k(z,v)$ can be obtained in the form

$$G_k(z,v) = v[kA(z,v) + B(z,v)], \tag{5.118}$$

with $A(z,v)$ and $B(z,v)$ evaluated by substituting (5.118) in (5.117) with $k = 1, 2$. A long algebra gives

$$A(z,v) = \frac{vz(1-z)^3}{[1 - (4+v)z + (6+v)z^2 - (4-v+v^2)z^3 + (1-v)z^4]}, \tag{5.119}$$

$$B(z,v) = \frac{(1-z)^2[(1-z)^2 - vz]}{[1 - (4+v)z + (6+v)z^2 - (4-v+v^2)z^3 + (1-v)z^4]}. \tag{5.120}$$

Note that $A(z,1)$ and $B(z,1)$ reduce to $A(z)$ and $B(z)$ in (5.68).

The parallel cluster size measure, $r_{\parallel}(N)$, can also be defined as the first moment of the spanning size $L$, i.e.,

$$c_N r_{\parallel}(N) = \sum_{L=1}^{N} L c_N(L) \sim N^{\nu_{\parallel} - \theta} \lambda^N, \tag{5.121}$$

with

$$c_N(L) \equiv \sum_{k=1}^{N-(L-1)} c_{N,k}(L). \tag{5.122}$$

The appropriate generating function is defined as

$$H(z) \equiv \sum_{N=1}^{\infty} [c_N r_{\parallel}(N)] z^N = \sum_{N=1}^{\infty} \sum_{L=1}^{N} \sum_{k=1}^{N-(L-1)} L c_{N,k}(L) z^N. \tag{5.123}$$

This can be further rearranged to give

$$H(z) = \left[ \frac{\partial}{\partial v} \sum_{k=1}^{\infty} z^k G_k(z,v) \right]_{v=1} = \left[ \frac{\partial}{\partial v} A(z,v) \right]_{v=1}. \tag{5.124}$$

Using (5.119) we finally obtain

$$H(z) = \frac{z(1-z)^3(1-4z+6z^2-3z^3+z^4)}{(1-5z+7z^2-4z^3)^2}. \tag{5.125}$$

Since the asymptotic form in (5.121) corresponds to the $(1-\lambda z)^{-(\nu_{\parallel}-\theta+1)}$ singularity in $H(z)$, the result (5.125) confirms $\nu_{\parallel} = 1$.

We can now address the question of finite-size properties. Observe that $A(z,v)$, given by (5.119), can be represented as

$$A(z,v) \equiv \sum_{k=1}^{\infty} z^k G_k(z,v) = \sum_{L=1}^{\infty} v^L \sum_{N=L}^{\infty} c_N(L) z^N, \tag{5.126}$$

with $c_N(L)$ given by (5.122). If we regard the $z^N$ factor as fugacity in the grand canonical ensemble, then $A(z,v)$ generates the fixed-$L$ partition functions

$$Z_L(z) \equiv \sum_{N=L}^{\infty} c_N(L) z^N, \tag{5.127}$$

which can be used in calculating thermodynamic quantities. Also, one possible definition of parallel correlation length is

$$\xi_{\|}(z) = \left[\frac{\partial \ln A(z,v)}{\partial v}\right]_{v=1} = \frac{H(z)}{A(z,1)} \sim (z_c - z)^{-1}, \tag{5.128}$$

where the asymptotic divergence with $\nu_{\|} = 1$ follows from the explicit results derived earlier.

Consider now the system of *finite* extent $M$ along the $X$ axis, i.e., the $X$ coordinates of the cluster sites are restricted to $0, 1, ..., M - 1$. The appropriate generating function is analogous to $A(z,v)$ in (5.126), but with $L$ values restricted to $L \leq M$,

$$A_M(z,v) \equiv \sum_{L=1}^{M} v^L Z_L(z). \tag{5.129}$$

In analogy with (2.53) one can define a scaling function, $P(\zeta)$, via

$$\frac{A_M(z,1)}{A_\infty(z,1)} \approx P(\zeta), \tag{5.130}$$

where the scaling combination $\zeta$ is defined by

$$\zeta \equiv M/\xi_{\|}(z) \sim (z_c - z)M. \tag{5.131}$$

Similarly, in analogy with (5.128), we can define the finite-size correlation length

$$\xi_M(z) \equiv \left[\frac{\partial \ln A_M(z,v)}{\partial v}\right]_{v=1}, \tag{5.132}$$

with the scaling behavior given by

$$\frac{\xi_M(z)}{\xi_{\|}(z)} \approx Q(\zeta). \tag{5.133}$$

Let us now consider the ratio (5.130). The $M = \infty$ generating function $A(z, v)$ given by (5.119) can be written in the form

$$A(z, v) = \frac{(1 - z)^3}{z^2}[v_-(z) - v_+(z)]^{-1}\left[\frac{1}{1 - v/v_-(z)} - \frac{1}{1 - v/v_+(z)}\right], \quad (5.134)$$

with

$$v_\pm(z) = -\frac{(1 - z)^2}{2z^2}[1 + z \pm \sqrt{1 + 6z + z^2}], \quad (5.135)$$

being the roots of the denominator of (5.119). The fixed-$L$ partition function $Z_L(z)$, defined by (5.129), is therefore given by

$$Z_L(z) = \frac{1 - z}{\sqrt{1 + 6z + z^2}}[v_-^{-L}(z) - v_+^{-L}(z)]. \quad (5.136)$$

With this result one can calculate exactly various finite-$M$ quantities. For the scaling analysis it is convenient to calculate the difference

$$A_\infty(z, 1) - A_M(z, 1) = \sum_{L=M+1}^{\infty} Z_L(z). \quad (5.137)$$

The functions $v_\pm(z)$ given by (5.135), have the following property for $0 < z < z_c$:

$$-v_+(z) > v_-(z) > 1; \quad (5.138)$$

while at $z_c$, we have $-v_+(z_c) > v_-(z_c) = 1$. For large $M$ and $L$ the contribution to (5.137) due to the $v_+(z)$ term in (5.136) is exponentially small. The contribution from $v_-(z)$ diverges as $z \to z_c^-$. Thus, the difference in (5.137) becomes

$$A_\infty(z, 1) - A_M(z, 1) \approx \frac{1 - z}{\sqrt{1 + 6z + z^2}}\frac{v_-^{-M}(z)}{v_-(z) - 1}. \quad (5.139)$$

In the limit $z \to z_c^-$, the denominator in this expression vanishes, while $v_-^{-M}(z)$ can be expressed as

$$v_-^{-M}(z) = \exp[-M \ln v_-(z)] \approx e^{-k\zeta}. \qquad (5.140)$$

The constant $k$ can be obtained explicitly: after a long calculation one gets $k \equiv 1$. Using this result, we finally obtain the scaling function

$$P(\zeta) = 1 - Ke^{-\zeta}, \qquad (5.141)$$

with $K \equiv 1$ as $z \to z_c$ (this result also requires a lengthy calculation). In deriving (5.141) we used (5.139)-(5.140) and the definition (5.130). Thus, the finite-size scaling function,

$$P(\zeta) = 1 - e^{-\zeta}, \qquad (5.142)$$

is obtained. [Similarly, one can calculate the scaling function $Q(\zeta)$ defined by (5.133). We omit the details of this calculation.]

## G. Selected literature

For a general survey of difference equations and "Ramanujan's Mathematics" one can consult the following literature.

**V. Privman** and **N.M. Švrakić** (1988) J. Statist. Phys. **51**, 1091

**C. Adiga, B.C. Berndt, S. Bhargava** and **G.N. Watson** (1985)

Mem. Amer. Math. Soc. **53** [315], 1

**G.E. Andrews** (1976) *The Theory of Partitions* (Addison-Wesley, Reading)

**W. Gautschi** (1967) SIAM Rev. **9**, 24

Stacking and compact animal models have been reviewed by the present authors (the first reference above): Our presentation in this chapter followed closely that review. The stacking models have been introduced by H.N.V. Temperley, Proc. Camb. Phil. Soc. **48**, 683 (1952), and further studied by F.C. Auluck, Proc. Camb. Phil. Soc. **47**, 679 (1951); H.N.V. Temperley, Phys. Rev. **103**, 1 (1956); B. Derrida and J.P. Nadal, J. Physique (Paris) Lett. **45**, L-701 (1984). Models of compact animals and stacks, and connections with difference equations, have been explored recently by V.K. Bhat, H.L. Bhan and Y. Singh, J. Phys. **A19**, 3261 (1986); V. Privman and G. Forgacs, J. Phys. **A20**, L543 (1987); M.L. Glasser, V. Privman and N.M. Švrakić, J. Phys. **A20**, L1275 (1987); V. Privman and N.M. Švrakić, Phys. Rev. Lett. **60**, 1107 (1988). Finite-size scaling properties were studied by G. Forgacs and V. Privman, J. Statist. Phys. **49**, 1165 (1987).

# VI. SUMMARY

In this work we have considered three classes of lattice models: walks, interfaces, and clusters, both at surfaces and in the bulk. Restrictions of directedness and compactness have rendered many of these models exactly solvable, some in full detail including scaling and finite-size behavior. Our goal has been to present these solutions and emphasize the resulting scaling properties.

In Chapter II we have considered directed walk models for conformational properties of linear polymers. These models have been solved in general dimensionality $d$, and their scaling properties analyzed exactly. We thus feel that this class of problems is reasonably well understood. Of course, from the point of view of exact solutions the *isotropic* self-avoiding walk models for $d \geq 2$ remain a major challenge.

Chapter III is devoted to two dimensional SOS models of interfaces, and our analysis has been focussed on the wetting transition. This model has been solved exactly only for a limited number of external (substrate) potentials including short range, and $1/r$ long range potentials. Recall that the RSOS model with the $1/r$ potential exhibits unusual features such as a nonscaling shift of the critical (wetting) temperature and, for the appropriate values of potential parameters, undergoes a first-order transition with divergent correlation lengths. It would be valuable to have solutions of discrete models with other types of long range potentials. On the more technical side, the development of the mathematical theory of analytic properties of the continued fraction (3.40) would be highly desirable. With such a theory, the continued fraction method for solving equations like (3.5) would yield results of practical interest.

The directed polymer chain adsorption, studied in Chapter IV, is one of the few solvable models of surface effects in two and three dimensional systems. An important extension of this model would be to consider the behavior of the *collection* of chains near a surface.

Finally, in Chapter V, we have described exact solutions of several compact cluster models. In addition, we have analyzed finite-size properties of the partially directed compact lattice animal model. It would be useful to have a similar finite-size study performed for other models, particularly in the cases when the generating functions have essential singularities.

We emphasize that all the above models were considered on *regular* lattices. Recent results by several workers suggest that introduction of randomness or impurities of various types leads to dramatic changes in behavior. Finally, as a more general reminder, we mention that exact solutions of the *three-dimensional* SOS and cluster models are scarce.

# INDEX

# Lecture Notes in Mathematics

# Lecture Notes in Physics

# Springer Series in *Materials Science*

**Editors:** U. Gonser, A. Mooradian,
K. A. Müller, M. B. Panish, H. Sakaki
**Managing Editor:** H. K. V. Lotsch

Volume 4

**S. Sugano,** Tokyo; **Y. Nishina,** Sendai;
**S. Onishi,** Kanagawa, Japan (Eds.)

## Microclusters

Proceedings of the First NEC Symposium,
Hakone and Kawasaki, Japan,
October 20–23, 1986

1987. IX, 289 pp. 187 figs. ISBN 3-540-17675-6

Volume 6

**G. Benedek,** University of Milan, Italy;
**T. P. Martin,** Max-Planck-Institut für Fest-
körperforschung, Stuttgart, FRG;
**G. Pacchioni,** University of Milan, Italy (Eds.)
**J. P. Toennis** (Guest-Ed.)

## Elemental and Molecular Clusters

Proceedings of the International School, Erice,
Italy, July 1–15, 1987

1988. VIII, 377 pp. 218 figs. ISBN 3-540-19048-1

Springer-Verlag Berlin
Heidelberg New York London
Paris Tokyo Hong Kong

# Springer Series in Solid-State Sciences

**Editors:** M. Cardona, P. Fulde, K. von Klitzing,
H.-J. Queisser
**Managing Editor:** H. K. V. Lotsch

Volume 63

**H. Kuzmany,** University of Vienna, Austria;
**M. Mehring,** University of Stuttgart; **S. Roth,**
Stuttgart, FRG (Eds.)

## Electronic Properties of Polymers and Related Compounds

Proceedings of an International Winter School,
Kirchberg, Tirol, February 23–March 1, 1985

1985. XI, 354 pp. 267 figs. ISBN 3-540-15722-0

Volume 76

**H. Kuzmany,** University of Vienna, Austria;
**M. Mehring,** University of Stuttgart; **S. Roth,**
Stuttgart, FRG (Eds.)

## Electronic Properties of Conjugated Polymers

Proceedings of an International Winter School,
Kirchberg, Tirol, March 14–21, 1987

1987. XIII, 442 pp. 265 figs. ISBN 3-540-18582-8

*In preparation*
Volume 91

**K. Kuzmany,** University of Vienna, Austria;
**M. Mehring,** University of Stuttgart; **S. Roth,**
Stuttgart, FRG (Eds.)

## Electronic Properties of Polymers III

1989. Approx. 350 pp. ISBN 3-540-51319-1

Springer